Learning and Assessing

SCIENCE PROCESS SKILLS

Fourth Edition

Richard J. Rezba
Virginia Commonwealth University

Constance Sprague
Indiana University South Bend

Ronald Fiel
Morehead State University

KENDALL/HUNT PUBLISHING COMPANY
4050 Westmark Drive   Dubuque, Iowa 52002

Disclaimer

Adult supervision is required when students are working on science activities and projects. Use proper equipment (gloves, forceps, safety glasses, etc.) and take other safety precautions such as tying up loose hair and clothing and washing hands when the work is done. Use extra care with chemicals, dry ice, boiling water, or any heating elements. Hazardous chemicals and live cultures (organisms) must be handled and disposed of according to appropriate directions from teachers. Follow science fair's rules and regulations and the standard scientific practices and procedures required by schools. No responsibility is implied or taken for anyone who sustains injuries as a result of using the materials or ideas, or performing the procedures described in this book.

 Printed on recycled paper.

Book Team

Chairman and Chief Executive Officer	Mark C. Falb
Vice President, Director of National Book Program	Alfred C. Grisanti
Editorial Development Supervisor	Georgia Botsford
Developmental Editor	Tina Bower
Prepress Project Coordinator	Sheri Hosek
Prepress Editor	Angela Shaffer
Permissions Editor	Renae Heacock
Design Manager	Jodi Splinter

Cover background image © 2003 Goodshoot.

Printed in the United States of America
10 9 8 7 6 5 4 3 2 1

Dedication

To H. James Funk who is remembered for his humor, his love of science,
and his concern for our nation's children.

The goal setting activity on page xv is in keeping with Dr. Funk's high expectations
for those individuals whose chosen career is to teach children.

Author Information for Correspondence and Workshops

Dr. Richard J. Rezba
Professor
Virginia Commonwealth University
2301 Stemwell Boulevard
Richmond, VA 23236
E-mail: rjrezba@vcu.edu
804-745-4144

Constance Sprague
Lecturer in Science Education
Indiana University South Bend
1700 Mishawaka Avenue
South Bend, IN 46634-7111
E-mail: csprague@iusb.edu
574-237-4129

Ronald Fiel
E-mail: rfiel@mis.net

Contents

Preface

Teaching science is an awesome responsibility. Regardless of how you personally view science, the children you teach are depending on you to model good science and to teach them the skills they need to learn more about our increasingly scientific and technological world. *Learning and Assessing Science Process Skills* is designed to help you develop the knowledge and skills necessary to bring the science process skills to your students.

What are the science process skills? They are the things that scientists do when they study and investigate. Observing, measuring, inferring, and experimenting are among the thinking skills used by scientists or by you and your students when doing science. Much of the pleasure of both learning and teaching science is experiencing science. Mastering these process skills will help you develop the kind of science program that mirrors real science.

Organization

Learning and Assessing Science Process Skills is presented in two parts. In Part One you will learn and practice the skills of observing, communicating, classifying, measuring, inferring, and predicting. These skills are called The Basic Science Process Skills because they form the foundation for later and more complex thinking skills. Instruction on the Basic Science Process Skills begins in pre-school, is emphasized in the elementary grades, and continues into middle school and beyond. In Part Two you will learn the skills you and your students need to design and conduct scientific investigations in class and at home. These skills are known as The Integrated Science Process Skills because they are used together to do what many consider the ultimate in problem solving in science—*experimenting*.

Although Parts One and Two are primarily about helping you become *competent* in the science process skills, they are also about helping you become *confident* in your ability to help students learn the same skills. While you are learning these skills, you will also be learning *how* you learned them. In sections called, **Decision Making 1 and 2,** you will be asked to use what you learned to make instructional decisions about how process skill development can be enhanced in existing curriculum materials. Technology continues to provide us with new tools for instruction and we have highlighted several that are particularly useful in teaching science.

Features New to This Fourth Edition

- **Alignment to National Standards**—national science and mathematics standards that are met in each chapter are identified as (NSES) and (NCTM) for National Science Education Standards and National Council of Teachers of Mathematics, respectively.
- **State Science Standards**—related science standards selected from states across the nation are also included. Use www.eduhound.com/k12statetesting.cfm or a search engine to access additional state science standards.
- **Multiple-Choice Test Items**—released state test items of K–12 science were selected from various states and provided at the end of each chapter.

- **Terminology**—the terms used throughout the text have been aligned with both state and national standards. For example, the terms *independent and dependent variables* found in many state standards are now used as well as manipulated and responding variables.
- **Technology**—integration of instructional technology is an exciting feature new to Part One of this edition. Among those included are the use of computer microscopes and digital cameras.
- **Internet**—numerous helpful website addresses are included throughout the text. Every effort has been made to provide up-to-date addresses of these sites but they may change. Use a search engine described in the new technology feature of Part One to access additional websites.
- **Chapters**—have been reorganized and rewritten with new laboratory activities. Additional teaching suggestions have been added throughout the text.
- **Teaching**—new teaching suggestions have been added throughout the text. Classroom scenarios have been added to illustrate how process skills are infused into science instruction.

Although originally intended as a text for K–8 teachers to learn the skills necessary to develop a hands-on and minds-on science program, earlier editions soon served other needs as well. For more than two decades, this book has been a source of both instructional and assessment ideas for practicing classroom teachers and curriculum developers.

In this new edition, we have continued our focus on assessment by including national standards in science and mathematics in each of the 16 chapters. In addition we have selected sample state standards to provide insight into the kinds of expectations our states have for K–8 students in science. Because evaluation often sends a message of what is considered important, we have included throughout the book sample multiple-choice test items as well as a full array of other forms of assessment for the science process skills.

Technology continues to provide us with new tools for instruction and we have highlighted several that are particularly useful in teaching science.

Teaching is about decision making. An already overcrowded curriculum prevents the easy decision of just adding a new unit on measurement skills or another on identifying variables. A more important decision making skill involves the ability to recognize opportunities in existing curricula to enhance the teaching, learning, and assessing of the science process skills. In Decision Making 1 and 2 typical science activities are presented to illustrate how existing instructional materials can be modified to better emphasize the science process skills. Teaching tips have been added throughout the text to help you accentuate the process skills in your science program. Finally, a fresh new look to *Learning and Assessing Science Process Skills* not only is more user-friendly, but represents, more than ever, the timely significance of learning these science thinking skills.

A Challenge

With every adventure there lies a risk. The risk here is that the individual chapters of the text imply a separation of the process skills used to do science. In truth these thinking skills of science are interdependent. As you study the science process skills, we hope the artificial separations created here will dissolve and, by merging these skills you become more able and confident in providing exemplary science instruction for your students.

The Authors

Introduction

Opportunity, High Standards, and Accountability

The Industrial Age that helped make America great relied on our country's natural resources and the hard work of our people. Although a strong work ethic and the wise use of our natural resources continue to be important in the Information Age, working smarter not just harder will secure our position in an increasingly global marketplace. Major changes in our society and in the world are causing a rethinking of the purposes of education, especially in mathematics and science. The goals for science education in the new millennium and beyond stress science as ways of thinking and investigating as well as a body of knowledge.

Ways of thinking in science are called the process skills. When scientists and students do science they are using such thinking skills as inferring, classifying, hypothesizing, and experimenting. The science process skills, along with the knowledge those skills produce, and the scientific values and habits of mind define the nature of science. Unfortunately, the teaching and learning of science does not always reflect the true nature of science. Too often students are burdened with short-lived learning of facts and boldfaced terminology at the expense of ever actively doing science. An increasing body of research supports the notion that students learn best when actively engaged, both physically and mentally, in hands-on and minds-on activities. There is also growing agreement among teachers and policy makers that less is more, where focusing greater attention on fewer concepts and skills is far more beneficial to students than covering vast amounts of abstract science content. Hands-on activities leads to minds-on understanding when teachers can concentrate on a more manageable number of big ideas and where students have regular opportunities to think about what they have been doing.

During the nineties the National Science Education Standards Project sponsored by the National Academy of Science, Project 2061 developed by the American Association for the Advancement of Science (AAAS), and the Scope, Sequence, and Coordination Project of the National Science Teachers Association (NSTA) were all long term reform initiatives in science education that have extended into this new century. The standards, benchmarks, and guidelines of these national reform efforts uniformly emphasized the need to create learning environments that encourage students' understanding of the scientific endeavor as well as students' excitement and enjoyment in its pursuit. High standards characterized these efforts because low expectations for students lead to fact-driven memorization with few opportunities to explore and to experiment using the science process skills.

Unlike the curricular direction that followed the launching of the Russian Sputnik in 1957, there is now genuine commitment to science for all students, not just the elite. Earlier efforts were largely aimed at guaranteeing an adequate supply of scientists and engineers to meet our national needs. More recent efforts reject practices where populations of students defined by gender, race, ethnicity, physical disability, and economic status are excluded from opportunities to learn science and are discouraged from pursuing science as a career. There is general consensus that science belongs to everyone and that it is in our nation's best interest that all students become scientifically literate.

National Science Education Standards and AAAS' Project 2061 Benchmarks

As a result of activities in grades K-4, for example, all students should develop abilities necessary to do scientific inquiry. These include:

- Ask a question about objects, organisms, and events in the environment.
- Plan and conduct a simple investigation.
- Employ simple equipment and tools to gather data and extend the senses.
- Use data to construct a reasonable explanation.
- Communicate investigations and explanations.

K-4 Content Standard A: Science as Inquiry,
National Science Education Standards

By the end of the following grades, students should

2nd grade: Raise questions about the world around them and be willing to seek answers to some of them by making careful observations and trying things out.

5th grade: Offer reasons for their findings and consider reasons suggested by others.

8th grade: Know that hypotheses are valuable, even if they turn out not to be true, if they lead to fruitful investigations.

Benchmarks for Science Literacy, AAAS' Project 2061

All 50 states have embarked on education initiatives related to high standards and accountability. These efforts include the establishment of state-wide academic standards for all students and the assessments that measure student performance. Assessment is taking center stage as schools are held more accountable for what students know and are able to do. Most states now have defined consequences for unacceptable performance by students and their schools. At the federal level new education legislation called *No Child Left Behind* became law in 2002. This legislation requires each state to establish its own unique set of standards for reading, math, and science. Under *No Child Left Behind* legislation each state must establish its own annual tests aligned with state standards for grades three through eight to measure how successfully students are learning what is expected by the standards. These tests must yield specific objective data. Federal funds for education will only go to states that implement standards and tests to measure student performance related to those standards.

Assessment is the process of gathering evidence of students' abilities and achievements. It is a form of communication that provides the information so evaluations can be made. Assessments help teachers and parents determine what students know, are able to do, and what they still need to learn. Evidence about student performance in science is needed for a variety of purposes, such as making instructional decisions, tracking student progress, communicating judgments, and evaluating programs. For each of these purposes, data are collected from essentially three basic sources: observations, student responses to questions, and student products and performances. Science assessment should be about gathering evidence of students' achievements and then using that evidence to further the growth of science knowledge and skills for all students.

Assessment methods should allow students to demonstrate what they know and are able to do, not just what they do not know. Teachers and schools need to have a detailed level of assessment to truly understand student achievement. Because all assessments are at best estimates of what students know and are able to do there is a definite limit to the information that state tests alone can provide. Multiple forms of assessment are needed to supplement external statewide assessments because different types of assessment reveal different aspects of performance. Just as lecturing should not be the sole method of instruction, neither should multiple-choice tests be the only means of assessment. Multiple forms of assessment include open-response questions, performance tasks, portfolios, interviews, teacher rating forms, and a variety of checklists for teachers, students, parents, and peers.

One characteristic of some forms of assessment is their ability to connect assessment with instruction. Most current testing practices are only assessment tools, not teaching tools; assessment typically occurs only when instruction stops. It is interesting to note that assessment comes from the French assidere (Latin sedere) meaning to sit beside, suggesting that a much closer relationship should exist between instruction and assessment. A variety of methods is needed to bridge the gap between teaching and assessment. Assessments in science should allow students to use their science process skills and content knowledge from the science disciplines in much the same way as they do in science class. Assessment should mirror the science that is most important for students to learn.

Multiple forms of assessment are also consistent with what we know about learning. Students with differing learning styles should have varied opportunities to demonstrate what they know and are able to do in science. Assessment needs to facilitate each student's continued learning in science.

High standards, state standardized exams, and increasing accountability are among the challenges facing today's teachers. Teachers must balance these demands while providing interesting and varied instruction. Fortunately, there are superb teachers who systematically observe, challenge, and listen to students to lead the way in both teaching and assessing the performance of students. Their exemplary science classrooms are characterized by high expectations, challenging tasks, strong work ethic, mutual respect, and a belief in science for all students. Teachers are the richest sources of information about students. They have always been and will continue to be the major assessors of student achievement. Of the multiple ways to assess students, nine forms of assessment are described here and modeled in the 16 chapters of Learning and Assessing Science Process Skills (see Multiple Forms of Assessment on following page). Also included in each chapter are sample multiple choice items selected from released standardized tests from various states. These high stakes multiple-choice items are the actual questions students were given to measure their knowledge of the related science process skill. Classroom evaluation should provide students with experience in a full array of assessment procedures including multiple-choice test items.

Multiple Forms of Assessment

Assessment	Topic	Chapter
Open Response Question	Observing Keeping Factors Constant	1 12
Performance Task	Classifying Measuring Inferring Predicting Relationships Between Variables Defining Variables	3 4 5 6 10 14
Portfolio	Identifying Variables	7
Teacher Paper and Pencil Checklist	Communicating	2
Self/Peer/Family Checklist	Tabulating Data	8
Teacher Digital Assessment	Laboratory Behaviors	11
Teacher Rating Sheet	Graphing Skills Experimenting	9 16
Individual Performance within a Group Rating Form	Designing Investigations	15
Interview	Constructing Hypotheses	13

How to Use This Book

Much of what you learn from this book depends on how you use it. Begin each chapter by first reading the purpose and objectives; knowing what is expected of you will help you learn the skills presented in the chapter. Do all of the activities because mastering the science process skills can only be achieved by being actively involved. Most activities have self-check sections that provide you with feedback on your responses. The special symbol, ✔, will cue you that the answers to the self-check questions will appear next. Write your responses before reading the answers. At the end of each chapter a Self-Assessment is provided for you to demonstrate your knowledge and skills. Use the answers provided to check your level of mastery. Experience with the chapter self-checks, self-assessments, and student assessment examples will also help you consider how you will assess your own students' abilities to use the science process skills. As you learn the process skills, think about how that experience will help you decide how to teach these same skills to your students. Study the examples in the sections called **Decision Making 1 and 2** to help you enhance the teaching of process skills.

The activities in the book were designed for either individual or small group study. You are encouraged, however, to work cooperatively with at least one other person as you proceed through the chapters. Working together may help you better process information, practice skills, receive feedback, and have more fun learning.

If you are using this book in a course, your instructor may suggest different materials to use in class or at home as you complete the activities in this book. Using different materials may result in some of your answers being somewhat different from ours. Those of you who are inservice teachers using ideas and activities from this book with your students may wish to substitute one kind of material for another, such as plastic cups for beakers. In addition you may wish to make overhead transparencies of some pages for use in your instruction and to use assessment examples throughout the book as part of your assessment program.

References

Brown, J.H. and Shavelson, R. J. (1996). *A Teacher's Guide to Performance Assessment.* Thousand Oaks, CA: Corwin Press, Inc.

Cothron, J, Giese, R, and Rezba, R. (2000). *Students and Research: Practical Strategies for Science Classrooms and Competitions,* 3rd ed. Dubuque, IA: Kendall/Hunt Publishing Company.

Doran, R., Chan, F., Tamir, P., and Lenhardt, C. (2002). *Science Educator's Guide to Laboratory Assessment 2nd ed.* Washington, DC: National Science Teachers Association Press.

Enger, S. K. and Yager, R. E. (2000). *Assessing Student Understanding in Science–A Standards-Based K-12 Handbook.* Thousand Oaks, CA: Corwin Press, Inc.

Hart, D. (1994). *Authentic Assessment: A Handbook for Educators.* New York: Addison-Wesley Publishing Company.

Hein, G. and Price, S. (1994). *Active Assessment for Active Science.* Portsmouth, NH: Heinemann.

Lowery, L.F. (1997). *NSTA Pathways to the Science Standards: Guidelines for Moving the Vision into Practice.* Washington, DC: National Science Teachers Association

National Research Council (2001). *Classroom Assessment and the National Science Education Standards.* Washington, DC: National Academy Press.

National Research Council (1996). *National Science Education Standards.* Washington, DC: National Academy Press.

National Science Teachers Association (1992). *Scope, Sequence and Coordination Content Core, Volume I.* Washington, DC: National Science Teachers Association.

National Science Teachers Association (1992). *Scope, Sequence and Coordination of Secondary School Science, Relevant Research, Volume II.* Washington, DC: National Science Teachers Association.

NWREL Resource Library (1998). *Bibliography of Assessment Alternatives: Science.* Portland, OR: Northwest Regional Educational Laboratory.

Pellegrino, J., Chudowsky, N., and Glaser, R. eds. (2001). *Knowing What Students Know: The Science and Design of Educational Assessment.* Washington, DC: National Academy Press.

Project 2061, American Association for the Advancement of Science (1993). *Benchmarks for Science Literacy.* New York: Oxford University Press, Inc.

Setting Goals

Children benefit most from those teachers who set goals for themselves. Think about what you would like to achieve in your first three years of teaching science. Write the first few sentences of a newspaper article that describe what you would like written about the way you teach science to elementary and middle school students.

Note: Write the newspaper article in pencil. After you have completed this book you will be asked to return to this article to make any changes you wish that reflect changes in your goals.

The World Reporter

Volume CXVII

Teacher Achieves 3-Year Goal

Part 1

Basic Science Process Skills

The *basic science process skills* are what people do when they *do science*. Children using these same skills are active learners. They use their senses to *observe* objects and events and they look for patterns in those observations. They *classify* to form new concepts by searching for similarities and differences. Orally and in writing, they *communicate* what they know and are able to do. To quantify descriptions of objects and events, they *measure*. They *infer* explanations and willingly change their inferences as new information becomes available. And they *predict* possible outcomes before they are actually observed. Learning science this way may be very different from the way you learned science in your elementary and middle school experiences. For you to teach the science process skills to children and to be able to implement a science curriculum that emphasizes these skills, you must first learn them yourself.

Experimenting

Describing Relationships Between Variables

Analyzing Investigations

Designing Investigations

Constructing a Graph

Acquiring and Processing Data

Constructing a Table of Data

Constructing Hypotheses

Defining Variables Operationally

Identifying Variables

Communicating

Predicting

Classifying

Measuring

Inferring

Observing

While learning the basic science process skills, you too will be an active learner. The activities in Part One have been carefully designed to help you focus on the process skills, and all of them have been successfully used in elementary and middle school classrooms. By thinking about how these activities helped you learn these skills, you can look for similar activities to help your students learn the process skills in much the same way.

The activities in this book use simple, ordinary supplies. Good science can be learned with everyday materials; elaborate and costly equipment is not required. In fact, it is the very ordinary that often stimulates students to ask questions that lead them to fruitful inquiry.

1

You will begin Part One with the skill of *observing*. This is the science process skill on which all the others are based. Each time you learn a new skill, ask yourself these two questions:

Teaching Children

How am I learning this skill?
How will I teach this skill to students?

Answering these questions will help you think about teaching as well as learning the process skills. After you have completed Part One, Basic Science Process Skills, you will be asked to make some instructional decisions about how you might teach these same skills to children.

CHAPTER 1

Observing

National and State Standards Connections

- Develop an understanding of properties of objects and materials. (NSES K–4)
- Describe qualitative and quantitative change. (NCTM, Pre-K–2)
- Keep a journal record of observations, recognizing patterns of observations and summarize findings. (New Jersey Core Curriculum Content Standards: 5.2)

MATERIALS NEEDED

- ✔ a plant to observe
- ✔ magnifying lens, or jeweler's loupe*
- ✔ optical or digital microscope (optional)
- ✔ cornstarch
- ✔ mixing bowl or similar container
- ✔ spoon
- ✔ water
- ✔ sealable plastic sandwich bag
- ✔ a 35 mm film canister or a graduated cylinder
- ✔ baking soda

- ✔ vinegar
- ✔ paper towels
- ✔ tray or open container to catch spills
- ✔ gobstopper candies (or M&Ms)
- ✔ round container (225 or 500mL) The clear or transparent deli containers are perfect for this and many other activities.
- ✔ a metric ruler
- ✔ an equal arm balance or other sensitive scale

B. C. by permission of Johnny Hart and Field Enterprises, Inc.

*Don't know what a jeweler's loupe is? You can find a description of a jeweler's loupe at this site: http://www.the-private-eye.com Look under "materials."

Classroom Scenario

Mrs. B, now a veteran teacher, reflecting on her first few years of teaching:

I remember referring to children's minds as little hovercrafts. . . . they were all over the place, never really landing anywhere. I couldn't get my students to focus on much of anything. One year, when we studied ants as part of an insect unit, we went outside to look at ants in their own environment, we read stories about ants, looked at pictures of ants, and talked about ant societies. At the end of our study I asked students to draw a picture of ants living in their natural environment. Looking at some of their pictures, I couldn't tell whether they were ants or rabbits! Artistic abilities aside, I could tell students really didn't know how many body parts or how many legs ants have or even if they have antennae. Some even drew ears and fuzzy tails. Just looking at ants hadn't helped the children learn much about them. Then it hit me, I was the one 'hovering'. I just assumed that when we spent time looking at something, children saw and remembered what I wanted them to see and remember. It hadn't occurred to me to teach students how to observe, and to check that their observations are good ones. Now I work hard on building and improving children's observation skills.

Here's a scenario from Mrs. B's insect study this year:

Mrs. B: This week, class, we will be studying about insects. I want you to know what some insects look like, why they behave the way they do, and why they are important.

Kathy: Do you mean 'bugs'? I don't like bugs.

Mrs. B: Yes, Kathy, we sometimes call insects bugs. I think you'll like insects when you learn more about them. Insects are very important to us. Humans could not exist without them. Let's begin with an insect you know, ants. On your paper, draw a picture of an ant. And while you are making your drawings I want you to ask questions about ants. I'll write down your questions.

Here are some questions children asked:

"Does my ant have to be black?"

"How many legs does it have?"

"Where do the legs go?"

"Does it have eyes?"

"Who cares about ants?"

"Do they have ears?"

"Ants can bite you. Do they have teeth?"

"Do they have feet or just legs?"

"Where do they live?"

Mrs. B: For now let's look at just those questions that can be answered by making good observations. Each group will have a clear plastic container with an ant inside. Please keep the lid on. Ants are living things so be gentle with them.

Observe the ant through the plastic and use your magnifying lenses to make the ant look bigger. I want you to draw another picture that looks as much like a real ant as possible. Record in your drawings as many ant parts as you can. Let's go over your questions one by one to remind us what to look for. When we're done we'll return the ants to the exact place outside where we found them. They'll use their sense of smell outside to find their way home.

Kathy: Huh? You mean ants can smell stuff?

Goals

The purpose of these exercises is to help you to further develop your skills of observation and to learn about different kinds of observations you can make about your own environment.

Before you proceed, use the space in the box to make your own drawing of an ant from memory. While you are drawing, list at least three questions you have about what an ant really looks like.

Ant Drawing	Ant Questions
	1. _____
	2. _____
	3. _____

—— Classroom Scenerio ——

Performance Objectives

After completing this set of activities you should:

1. given an object, substance, or event, be able to construct a list of qualitative and quantitative observations about an object, substance, or event. Your observations must be made using at least four of your senses.
2. given an event in which a change is involved, be able to construct a list of qualitative and quantitative observations about the changes *before, during,* and *after* they occur.

To observe an object or substance means to carefully explore all of its properties. Objects may have such properties as color, texture, odor, shape, weight, volume, or temperature. They may even make sounds either on their own or when manipulated.

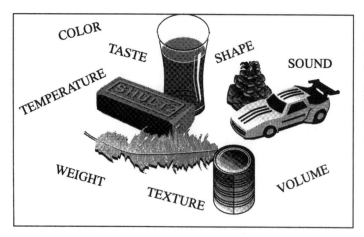

Different objects or substances have different properties. That's what makes them different from all other objects or substances. Through the use of our senses we are able to perceive an object's characteristic properties by seeing, hearing, touching, tasting, or smelling them. Observing involves identifying and describing an object's properties.

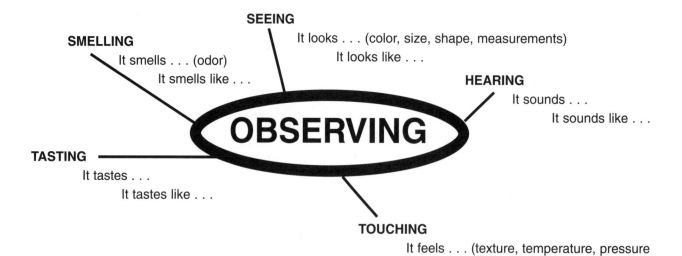

SEEING
It looks . . . (color, size, shape, measurements)
It looks like . . .

SMELLING
It smells . . . (odor)
It smells like . . .

HEARING
It sounds . . .
It sounds like . . .

OBSERVING

TASTING
It tastes . . .
It tastes like . . .

TOUCHING
It feels . . . (texture, temperature, pressure
It feels like . . .

Activity 1.1 Using All Your Senses

→**GO TO** one of the plants in the room, or at home, and gather as much information as you can about the plant using all your senses except taste. (CAUTION: Tasting any unknown substance is hazardous business; never taste anything unless you are absolutely certain that there is no danger involved.) In the chart following, list at least ten observations about the plant. For each observation record the sense you used to obtain the information.

Cover the answers in the self-check and refer to them only after completing the activity.

Observations	Senses
1. _____	_____
2. _____	_____
3. _____	_____
4. _____	_____
5. _____	_____
6. _____	_____
7. _____	_____
8. _____	_____
9. _____	_____
10. _____	_____

Wait, you are not done with your plant yet. ➡**GO TO** the supply area and pick up a magnifying lens or microscope, or both.

When using magnifying lenses for the first time with young children, you'll have to teach them how to focus them. The easiest method is for a child to hold the lens against his eye, and then slowly move his head toward the object until it can be seen clearly. Or alternately, move the object toward the eye. Think about how jewelers and Sherlock Holmes use a magnifying glass. Moving a magnifying lens back and forth over an object is much harder for children even though most adults use lenses this way.

Use the magnifier or microscope to extend your sense of sight. Do your observing in an area where there is plenty of light available. Record at least four additional observations that you were able to make using these tools.

Observations Made Using Lenses

1. _____

2. _____

3. _____

4. _____

Compare your observations with someone else's or with those on the following page.

Self-Check _____ Activity 1.1

Your list of observations should provide at least enough information to answer these questions about the plant you observed:

1. What color is it? Is the color evenly distributed? (sight)
2. Is the plant tall, short, spindly, sprawling? (sight)
3. Is there one main stem or many? (sight)
4. What is the general shape of the leaves? (sight)
5. Do the leaves have jagged or smooth edges? (sight)
6. Are the leaves shiny or dull? (sight, touch)
7. Are the leaves opposite one another or alternate? (sight)
8. Are the veins of the leaves distinct? Is there a central vein? Are the veins opposite one another or alternate? (sight)
9. Is the stem thick or thin? (sight, touch)
10. Are the leaves in clusters or separate? (sight)
11. What is the texture of the stem and leaf surfaces? (touch)
12. Do the leaves feel waxy? (touch)
13. Are the leaves stiff or easily pliable? (touch)
14. Does any part of the plant have an odor? (smell)

The additional observations you were able to make when using a magnifying lens or microscope should help you to answer these questions about the plant you observed:

1. Are there hairs on the top side of the leaf? On the bottom side?
2. Are the leaf edges as smooth or jagged as you first observed?
3. Are there any holes, cracks or imperfections on the leaf's surface that you did not see before?
4. Are there colors you did not see before?
5. Are there any other structures you did not see before (bumps, spines, ridges)?

Activity 1.2   Making Qualitative and Quantitative Observations

Most of your observations in Activity 1 were probably *qualitative* observations; that is, observations you made that were about the plant's characteristics or qualities. The following statements are examples of qualitative observations that you may have made about the plant:

- It is light green in color. (sight)
- It has a pungent odor. (smell)
- Its leaves are waxy and smooth. (touch)
- It makes a rustling sound when lightly rubbed. (hearing)

When we want more precise information than our senses alone can give us, we include a reference to some unit of measure. Measurements could be made using a standardized unit like centimeters or inches, or non-standard units like paperclips as units of length. Observations that provide information about quantity, like a number or amount, are called *quantitative* observations. Quantitative observations help us communicate specifics to others and provide a basis for comparisons. The following statements are examples of quantitative observations that could be made about a plant.

- One leaf is 10 cm long and 6 cm wide. (metric ruler)
- The mass of one leaf is 5g. (balance)
- The temperature of the room in which it grows is 22°C. (thermometer)
- This plant's leaves are clustered in groups of five.
- This plant is larger than that plant.
- Each flower is as wide as 3 paper clips placed end to end.

Quantitative observations made with standardized instruments such as rulers, meter sticks, balances, and graduated cylinders, give us very specific and precise information. Although approximations and comparisons, such as larger, shorter, and heavier are not precise, they are also considered quantitative observations.

In this activity, rather than observing a plant, you will observe an unusual substance. First, you need to follow a recipe to make the substance, which is called a non-Newtonian fluid—one that does not follow all the conventional expectations for liquids. Then you will practice making qualitative and quantitative observations about the substance you made. To find recipes for other non-Newtonian fluids on the Internet, try the search phrase "recipes for non-Newtonian fluids."

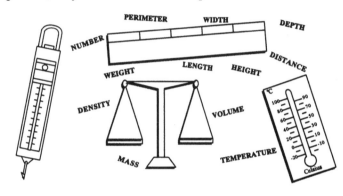

➔**GO TO** the supply area and obtain:

a bowl or other container for mixing,

a spoon or popsicle stick for stirring

75 ml of cornstarch (about the amount in a 3-ounce paper cup)

35 ml of water

Procedure

1. Slowly add the 75 ml of cornstarch to the 35 ml of water, stirring constantly.
2. You may have to adjust the amounts a little so the substance is like the consistency of toothpaste, not too dry or too runny.
3. Ask your instructor how to dispose of the substance when you finish.

Now that your substance is ready, formulate some questions to guide your observations just as you would for your students, such as, *Can you pinch it? Can you pick it up? Can you shape it? Can you poke it? Is it cold? Does it have an odor?* In the charts that follow, list at least five qualitative observations and four quantitative observations about the substance. For each qualitative observation, identify the sense you used to gain the information and for each quantitative observation identify the instrument you used to aid your senses.

Qualitative

Observations	Sense Used
1. _____	_____
2. _____	_____
3. _____	_____
4. _____	_____
5. _____	_____

Quantitative

Observations	Measurement Tool Used
1. _____	_____
2. _____	_____
3. _____	_____
4. _____	_____

Compare your answers with someone else's or check your answers with those that follow.

Self-Check ———————————————————————————————— Activity 1.2

Some of the observations you may have made are included in the following chart. Of course there are other acceptable observations. If you are in doubt about any of your observations, please ask a peer or your instructor.

Qualitative Observations	Sense Used
1. Has a creamy white color, shiny	sight
2. Is runny when poured slowly	sight
3. Smells like grass	smell
4. Is sticky when touched gently	touch
5. Feels hard when hit or poked	touch
6. Feels cold to the touch	touch

Quantitative Observations	Measurement Tool Used
1. Mass: about 100 g	balance
2. It takes about 15 seconds for all the substance to flow from one side of the bowl to the other.	timer
3. A 2 cm diameter ball made of the substance spreads out into a puddle 4 cm in diameter when left to sit on the table.	ruler
4. When I pinch the substance, it breaks into many small pieces.	

✔

Activity 1.3 **Observing Changes**

You will often observe objects or phenomena that undergo physical or chemical changes. Your observations will be either qualitative, in which you use your senses to obtain information, or quantitative, in which you make a reference to some standard unit of measure. When asked to describe a change, it is important to include statements of observation made before, during, and after the change occurs.

Think, for example, about the changes you might observe when you make popcorn. Before the kernel is heated, it is teardrop shaped, about 1 cm x 0.5 cm x 0.5 cm in size, light brown in color, and has a hard, smooth shell. During the change (popping) the shell splits, a white puffy mass expands through the shell, and a short, light sound is produced. After the change the piece of popcorn is irregular in shape, about 3 cm x 2 cm x 3 cm in size, has a white, puffy texture, and a corn-like taste. Of course, more observations could be made.

You will often observe objects or phenomena as they undergo physical or chemical changes. Your observations will be either qualitative, in which you use your senses to obtain information about characteristics or qualities, or quantitative, in which you typically make reference to some unit of measurement. When asked to describe a change, it is important to include statements of observations made before, during and after the change occurs.

In this activity you will create a chemical reaction.and observe what happens *before, during* and *after* the reaction.

→GO TO the supply area and pick up the following items:

a *sealable* plastic sandwich bag

It is very important that the sandwich bag has no holes. Fill the bag about 1/4 full of water, zip the bag closed, turn it upside down and check for leaks. When you find a bag that does not leak, that is the one you will use. Unzip the bag and pour out the water.

water for testing the sandwich bag for leaks

a 35 mm film canister

baking soda, enough to fill a film canister (about 35 mL)

vinegar, enough to fill a film canister (about 35 mL)

paper towels for cleaning up any spills

Mess alert: Just in case things do not go entirely as planned, conduct this activity on a tray or in an open container to catch spills.

Procedure

1. Fill a film canister with vinegar (if you prefer to measure using a graduated cylinder, this is about 35 mL) and pour the vinegar into a sealable plastic sandwich bag.

2. Fill the film canister with baking soda and carefully stand the *upright* canister into the sandwich bag that you have placed on a tray or in a container. At this point you want to keep the vinegar and baking soda separate. *Zip the bag closed.* See the drawing.

3. Make some *before* observations, both qualitative and quantitative. Record your observations in the chart.

Qualitative Observations	Quantitative Observations
Before	
1. _____	1. _____
2. _____	2. _____
3. _____	3. _____
4. _____	4. _____
5. _____	5. _____
During	
1. _____	1. _____
2. _____	2. _____
3. _____	3. _____
4. _____	4. _____
After	
1. _____	1. _____
2. _____	2. _____
3. _____	3. _____
4. _____	4. _____
5. _____	5. _____

4. Turn the bag so the baking soda and vinegar mix together, and place it back on the tray or in an open container.

5. Now, make some *during* observations, both qualitative and quantitative. Record your observations in the chart.
6. After the fizzing stops, make several *after* observations, both qualitative and quantitative. Record your observations in the chart.
7. Be sure you rinse and dry the film canister. Bags could also be rinsed and turned upside down to dry.

Compare your answers with someone else's or check your answers with the ones that follow. Then proceed to the Self-Assessment.

Self-Check _____ Activity 1.3 ✔

Below is a list of observations that could be made about the reaction between baking soda and vinegar *before, during* and *after* the reaction has occurred. Some of your observations may be different depending upon the actual volume of substances you use, how you mix them together, and the availability of a sensitive scale. This, of course, is not a complete list. Please see your instructor if you have questions about the observations you have made.

Qualitative	Quantitative
Before	
baking soda:	
1. white color	1. volume: one film canister full (35 ml)
2. feels dry, soft and powdery with very fine particles	2. mass: 32 grams
3. can be compressed slightly but doesn't hold its shape well when pinched, sometimes forms little irregularly shaped clumps	3. when clumps occur they are about the size of a pea or smaller
4. powder "squeaks" when you rub it between your fingers	4. smallest particles are the size of fine dust
5. has a faintly mild fragrance	5. baking soda is at room temperature
vinegar:	
1. clear, like water	1. volume: one film canister full (35 ml)
2. feels wet like water	2. mass: 34 grams
3. tastes sour	3. vinegar is at room temperature
4. smells like pickles	

Qualitative	Quantitative
bag: clear, shiny, flexible, flat	volume: about 14 cm × 14 cm × 1 mm
During	
1. fizzing sound occurs	1. the volume of the bag is expanding
2. white foam billows up in the bag	2. bubbles forming range in size from very small to about the size of a dime
3. bubbles form and pop	3. foaming bubbles fill nearly half the bag
4. bag puffs up and offers resistance to being squeezed	
After	
1. a white, milky liquid remains	1. volume of the expanded bag is 13 cm × 13 cm × 7 cm
2. if left to settle, layers form with a clear, watery layer on top and a white cloudy layer (looks like toothpaste) on the bottom	2. volume of liquid: a little less than 1 1/2 film canisters (about 50 ml)
3. no change in odor	3. total mass: 64 grams
4. feels cool to the touch	4. bag feels cooler than room temperature

✔

 # Self-Assessment   Observing

➔**GO TO** the supply area and obtain a clear round container, such as a deli container, half-filled with room temperature water. Also pick up 4 different colored Gobstoppers[1] candies. (You can substitute M&M's, but the results are more surprising with Gobstoppers.) In this assessment, you will be observing changes that take place as the Gobstoppers slowly dissolve in undisturbed water.

1. Before placing the Gobstoppers in the water make 3 qualitative and 2 quantitative observations.
2. Place the different colored Gobstoppers in the water, equally spaced around the edge of the container. You will need to observe them for several minutes.

---

[1] Gobstoppers are available wherever bulk candy is sold. They are also sold in packages in many national chain stores, such as 7-Eleven and CVS. Or use a search engine to find numerous Internet sources.

3. List at least three qualitative observations *while* changes are occurring.

4. After two (or more) color changes, record what you observed. At least two of your observations should be quantitative. You may wish to remove the Gobstoppers from the water so you can make additional observations.

Compare your answers with someone else's. Be sure that your observations cover the properties listed in the Self-Assessment answers.

Qualitative Observations	Quantitative Observations
Before	
1. _____	1. _____
2. _____	2. _____
3. _____	
During	
1. _____	
2. _____	
3. _____	
After	
1. _____	1. _____
2. _____	2. _____
3. _____	

Self-Assessment Answers

The observations you made about Gobstoppers *before, during,* and *after* dissolving may have included statements about the following:

Qualitative	**Quantitative**
Shape	Number of color changes
Color	Change in diameter
Texture	Change in mass
Smell	Time it took to change from one color to the next
Taste	Distance the colors diffused

IDEAS FOR YOUR CLASSROOM

1. Objects that can be interesting to observe are flowers, fruits, a pine cone, different kinds of leaves, feathers, and dried foods such as cereals.

2. Events such as popcorn popping, making ice cream, making butter or cookies can be delicious as well as informative.

3. A simple drop of water can be fascinating and lead to many challenging questions. Place a single drop of water on a paper towel or ordinary paper. What happens? (Water is attracted by paper fibers and is absorbed.) Place a drop of water on waxed paper. What happens? (The water drop "balls up"—cohesion.) Tip the waxed paper so that the drop moves. Does it roll or slide? How can you find out? (Hint: Sprinkle it with pepper or chalk dust.) Place a drop of water on plastic wrap. Place the plastic wrap so you can look at some printed material by peering through the water drop. What happens? (It magnifies.) Experiment with larger and smaller water drops to see which makes better magnifiers.

4. Observation helps us learn that important changes are taking place.
    a. Seal one nail in a plastic sandwich bag and another nail in a plastic sandwich bag with a dampened paper towel. Observe for several days. What differences were observed? Why? An interesting spin-off of this lesson would be to come up with different ways to prevent the nail from rusting.
    b. A similar lesson could be conducted using bread. Students would learn that moisture is important for rotting to take place.
    c. Observing changes taking place with a banana peel in a sealed sandwich bag could lead to interest in how different kinds of foods are preserved. It could lead to an interesting history lesson. How did pioneers preserve food? What did those foods taste like?

5. "Our Senses Depend on Each Other" is an enjoyable lesson. Have the students close their eyes so they aren't peeking and hold their noses so they can't smell. Give them a small sliver of apple, then raw potato, then raw onion. Can the students taste the difference? Now let them do it with their noses opened. Now can they tell the difference? Why does the sense of smell help the sense of taste?

6. "Autumn and the Five Senses" would make an excellent theme for a bulletin board and interest center. Fruits and vegetables, the changing color of leaves, and other changes could be observed.

7. "Safety and Our Senses" is another topic that is worthy of teaching. One way to approach this topic is to use sense deprivation. For example, tape could be placed on the ends of the fingers and students could compare how rough coarse sandpaper feels between taped and untaped fingers. This lesson could be extended by imagining no sense of touch. How would we be protected from sharp objects, heat, sharp blows, or blisters on our feet? Smell warns us when we are breathing something that could be harmful to us. You could use vinegar to simulate noxious gases. Smelling smoke can also warn children of the danger of fire and being burned. The sense of taste can be discussed as something that can warn of danger, but stress the hazards of tasting unknown substances. This would be a good time to discuss with students some of the poisons and poisonous plants found in the home. Sight and hearing are more obvious as signals of danger (e.g., horns, traffic lights, and so on), but students still need to learn about safety. Again, the lesson could be started by having students imagine how they would cope with danger if they had no sense of sight or hearing. The school nurse or a local doctor could be invited to talk to students about eye and ear care and perform sight and hearing tests.

What do you observe about that object (or event)?

List as many properties as you can about that object.

What are all the things you observe directly about . . . ?

Describe how this object looks, feels, smells, and sounds.

High Stakes Testing

A sample multiple-choice item from State Standardized Exams.

Which of these is the same as the animal above?

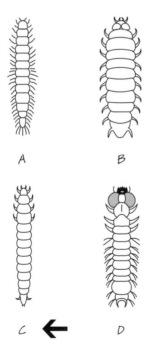

A   B

C ⬅   D

Virginia: SOL Grade 3 Spring 2001 Release Item.

http://mbgnet.mobot.org/pfg/cycles/
www.lessonplanspage.com/ScienceDiscoverTheWorldWithSenses13.htm
http://www.hhmi.org/coolscience/inchsquare/index.html
http://askeric.org/cgi-bin/lessons.cgi/Science/ProcessSkills
http://bugscope.beckman.uiuc.edu/
http://atm.geo.nsf.gov/instruction/observations.html

WEBSITES FOR ACTIVITIES

Technology Spotlight

Extending the Senses with a Digital Microscope

The Intel Play QX3 Computer microscope performs the functions of a traditional optical microscope with the advantages of digitized images. Although originally marketed as a toy, the QX3 has enormous instructional potential for K–12 classrooms. This versatile microscope can be used in its stand, or removed and held as a magnifying camera offering new options for student explorations. The magnifications are 10X, 60X, and 200X. Applications on the accompanying software offer numerous options not available with optical microscopes. Digital images can be stored, printed, and even assembled into a slide show. Other options include recording both real time and time-lapse movies, and customizing images using the Paint option to label and enhance images. In addition, all images and movies can be exported as jpeg images or avi movies for use in PowerPoint® presentations and on websites.

If a digital microscope is available, try looking at some common materials, such as salt, under the three magnification levels. Take slides of the salt at each magnification level. Remove the QX3 from its stand, and use it to explore your skin, clothing texture, and even your eye! If a digital microscope is not available, explore its use on the following website or use a search engine to locate additional sites:

www.micro.magnet.fsu.edu/optics/intelplay/index.html

Use the site's interactive Java tutorial—QX3 Microscope Simulator—to explore how the QX3 microscope software and hardware work together to capture images of a variety of specimens at different magnifications.

A Model for Assessing Student Learning

Assessment Type: Open-Ended Questions or Situation

Directions to the Students

A. Observe the aquarium[1].
B. Write three observations about the aquarium picture.

1. _____
2. _____
3. _____

Scoring Procedure

1 point for each correct observation.

Acceptable responses include:

■ Some fish are bigger than others.
■ There are two snails in the aquarium.
■ The aquarium is not completely full.
■ Snails are on the bottom of the aquarium.
■ The leaves of one plant look like feathers.

Unacceptable responses that are not observations include:

■ Snails and fish don't like each other.
■ More fish could live in the aquarium.
■ The small fish are babies of the big fish.
■ Fish near the top are breathing air.
■ The plants will not have enough air.

---

[1] Although an actual aquarium is preferred, a picture of an aquarium may be used. A terrarium or other interesting objects could be substituted as well.

From Rezba, Sprague, & Fiel. *Learning and Assessing Science Process Skills,* 4th Edition. © 2003 Kendall/Hunt Publishing Co. May be reproduced by individual teachers for classroom use only.

Chapter **2**

Communicating

National and State Standards Connections

- Develop the abilities to communicate investigations and explanations. (NSES K–4)
- Specify locations and describe spatial relationships. (NCTM Pre-K–2)
- Describe objects according to their physical properties. (Oregon Academic Content Standards: Benchmark 2, Grade 5)

MATERIALS NEEDED

To do the communicating activities in this chapter you will need:

✔ a set of 'Sensory Materials' (consisting of a variety of objects to smell, taste, feel, and see)

✔ a magnetic compass (optional)

 **Classroom Scenario**

It was a warm day and the children were anxious to go outdoors. Mrs. Q, their teacher, had prepared them well so they would know just what to do.

Mrs. Q: Before we go out into the schoolyard tell me again, what are your assigned roles?

Sam: I'm a materials manager and I get and return the stuff we need. And I get to say where our plot of land will be.

Mrs. Q: Right, each group will need a long piece of yarn, 4 stakes, a magnetic compass, a hand shovel, a tray and a magnifying glass.

Okra: What's the tray for?

Mrs. Q: You may want to put a soil sample or a creature you find on the tray so you can observe it better. But be sure to put everything right back where you find it as soon as you are through describing it.

Sally: I am a recorder and I draw the map and mark on it where we find things.

21

Mrs. Q: That's right and don't forget to show which way is north, south, east, and west on your map. That way you'll be able to talk about certain locations in your plot. Let me know if you need help using your compass.

Lori: I am a tracker. I make sure we do all the steps we're supposed to and fill in the chart.

Tam: I'm a checker. I make sure we all agree before we write stuff on the chart.

The children's chart looked like this:

What are some of its characteristics?	Is it living?	What do you think it is?	How many (or how much) did you find?	How do you think it got there?	What does it need to survive?

Mrs. Q: And what will you do with all the information you collect in your charts?

Brad: We'll use it to write a story to our *key pals* (the electronic equivalent of pen pals).

--- Classroom Scenario ---

The children in this scenario are learning and practicing their communication skills. They are collecting information, organizing it in a meaningful way, and communicating it to someone else. Did the lesson go as smoothly as Mrs. Q imagined? No. The children disagreed on just about everything, whether some things were living or not, whether 'up' was the same thing as north, and especially what their map was to look like.

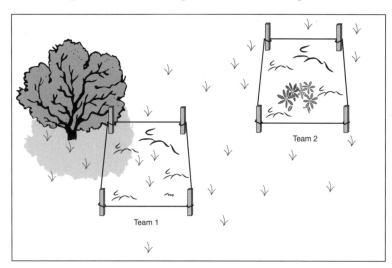

Team 2

Team 1

Mrs. Q was encouraged, however, by what her students were able to do. These children, who happened to live in Indiana, could not wait to send their e-mail messages to their *key pals* in Hawaii and to read about what the Hawaiian children found in their plots. Mrs. Q, of course, was sure to check that the stories contained only information gathered in the land plot investigations. One group wrote:

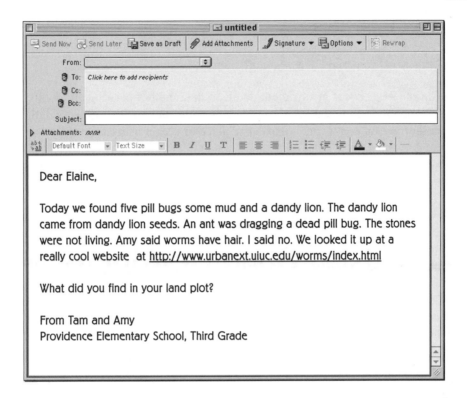

Dear Elaine,

Today we found five pill bugs some mud and a dandy lion. The dandy lion came from dandy lion seeds. An ant was dragging a dead pill bug. The stones were not living. Amy said worms have hair. I said no. We looked it up at a really cool website at http://www.urbanext.uiuc.edu/worms/index.html

What did you find in your land plot?

From Tam and Amy
Providence Elementary School, Third Grade

The class repeated the land plot investigation when weather conditions were different from the day of the first investigation. The children were surprised at how their charts had changed. The land plot investigation was just part of a thematic unit on communication Mrs. Q and other teachers at her grade level developed and called 'Get the story. Tell it well.' The goal was to get children involved in investigating objects, places and events, making good observations, and reporting what they find. The children had to first 'get the story' by investigating and gathering observable information, then organize their thinking and 'tell the story' to someone else. Other investigations included studying a pond, growing crystals, blowing soap bubbles, and comparing various kinds of glue. The teachers focused on getting children to stretch the number of ways they are able to communicate through the use of graphs, charts, maps, symbols, diagrams, mathematical equations, visual demonstrations, concept maps, rubrics, and the spoken word.

Effective communication is clear, precise, and unambiguous. It uses skills that need to be developed and practiced. The skills the children are practicing in the scenario are the same skills scientists use when they inquire, investigate, solve problems, collaborate with others, and report findings. When we hear or read the words, "studies show that. . . ." scientists are sharing what they have found.

New scientific studies seem to show that for the brain to function at its highest potential and for it to remain healthy throughout life, brain cell connections must be built early in life and then maintained throughout life. Of course, proper nutrition is important to this 'building' and 'maintaining' of healthy brains but early socialization and language building play a greater part than once thought. Teachers in Mrs. Q's school are collaborating not only on how to use this new information on brain research but also on how to share it with parents long before children begin attending school.

Goals

These exercises will help you learn to communicate ideas, directions, and descriptions effectively and give you practice in using and constructing various methods of communication.

Performance Objectives

After completing this set of materials, you should be able to:

1. Describe an object or event in sufficient detail so that another person can identify it.
2. Construct a map showing relative distances, positions, and sizes of objects with sufficient accuracy so that another person can locate a particular place or object using the map.

In the following activities you will build and practice your own communication skills. You will need to model them well for children.

Activity 2.1   Using Good Descriptors

Do this activity alone. Once you have completed it by yourself, talk with other students about what they observed about the objects and add to your list of possible descriptors.

➜**GO TO** the supply area and obtain a set of *Sensory Materials*. In this activity you will explore a wide variety of objects displaying several different properties. As you observe these objects think about how you would describe their properties to someone else. Your task is to generate a list of descriptive words (descriptors) that can be used to effectively communicate what you observe (smell, feel, taste, hear, and see) to others.

Keep in mind you are not attempting to name the objects or describe how you feel about the properties, you are just describing properties.

Some words to
describe how
things *smell*
_____   _____   _____
_____   _____   _____
_____   _____   _____

Some words to
describe how
things *feel*
_____   _____   _____
_____   _____   _____
_____   _____   _____

Some words to
describe how
things *taste*
_____   _____   _____
_____   _____   _____
_____   _____   _____

Some words to
describe how
things *sound*
_____   _____   _____
_____   _____   _____
_____   _____   _____

Some words to
describe how
things *look*
_____   _____   _____
_____   _____   _____
_____   _____   _____

Some possible descriptors are listed in the *Self-Check* that follows.

Self-Check _____ Activity 2.1

Some words to describe how things—

Smell   sweet, rotten, smoky, fresh, spicy, pungent, strong, moderate, weak, lemony, oily, minty, moldy, woody, sweaty, perfume-like.

Taste   sweet, sour, bitter, strong, moderate, weak, rich, spicy, syrupy, acidic.

Feel   rough, smooth, sandpaper, feathery, slick, cold, hot, warm, cool, rubbery, prickly, sharp, soft, hard, gritty, fuzzy, furry, scaly, cottony, bumpy, oily, waxy, sticky, wet, dry, moist, slippery, leathery, powdery, crumbly, creamy, glassy, jagged, slimy, vibrating.

Sound   loud, moderate, soft, brassy, high, low, medium pitch, sharp, dull, rattle, ringing, muffled, clear, distinct, squeaky, bark-like, scraping, tearing, banging, crashing, dripping, clicking, crinkling, abrupt, continuous, sudden.

Look   colors, shapes, designs, shiny, dull, clear, cloudy, sparkles, bubbly, bright, intense, continuous, interrupted, muted.

Communicating Descriptions

When you describe an object to someone, your purpose will be better served if your communication is an effective one. You can communicate effectively if you:

1. Describe only what you observe (see, smell, hear, and taste) rather than what you infer about the object or event.
2. Make your description brief by using precise language.
3. Communicate information accurately using as many qualitative and quantitative observations as the situation may call for.
4. Consider the point of view and past experience of the person with whom you are communicating.
5. Provide a means for getting *feedback* from the person with whom you are communicating in order to determine the effectiveness of your communication.
6. Construct an alternative description if necessary.

In the next activity you will communicate a description to someone else. You will then receive feedback on the effectiveness of your communication.

 Activity 2.2 ## Describing Objects to Others

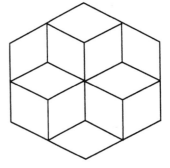

Look at the figure that follows:

Think about how you might describe this figure to someone in sufficient detail so that he or she could draw it from your description.

The artist will need to know what kind of lines to draw, where to place them, and how long they should be. Locate a person to be the artist and do the following:

- Look at the figure again and keep looking at it until you perceive it in a way that is different from how you first perceived it. (There are at least eight different ways to perceive this figure.) The way you describe something to someone else depends on how you perceive it.
- Carefully consider how you will describe the figure to the artist before you begin speaking.
- Without showing the figure to the artist, *effectively communicate* to that person how to make the lines so that their completed drawing looks as much like the original figure as possible.
- The similarity between the figures is a measure of the effectiveness of your communication.

Practicing Using Communication Tools

In the following activity you will describe an object to another person *without* telling that person what the object is. That person will then try to determine what the object is solely from your oral description of the object's properties. If your partner is able to identify the object, your communication must have been effective. Before you begin, familiarize yourself with the criteria listed in the accompanying chart. When you are through giving your description, your partner will use this chart to evaluate your communication.

 **Activity 2.3   Using a Rubric to Evaluate a Communication**

1. Choose a partner to work with you.
2. Select an object (or an organism) in the room to describe but do not tell your partner what the object is.
3. Plan your description in your head before you begin speaking.
4. Describe the object for your partner.
5. Have your partner try to determine what the object is from your description.
6. After the description is complete and the object has been identified, your partner should use 'Communicating Descriptions Rubric' to evaluate your description. A check in the 'Yes' column means that criteria was met. A check in the 'Needs Improvement' column means you need more practice.

Communicating Descriptions Rubric

Criteria	Yes	Needs Improvement
The Communicator ... 1. Describes only what is directly observed through the senses.		
2. Makes the description brief by using precise language.		

Criteria	Yes	Needs Improvement
3. Is accurate and uses sufficient qualitative and quantitative observations.		
4. Considers the point of view and experience of the receiver.		
5. Provides a means for getting feedback from the receiver about the effectiveness of the communication.		
6. Constructs an alternative description when necessary.		

Learning to use the *tools* of communicating helps children to be able to make good decisions about *how* to communicate observations and ideas. Here are some tools you can use to share what you know:

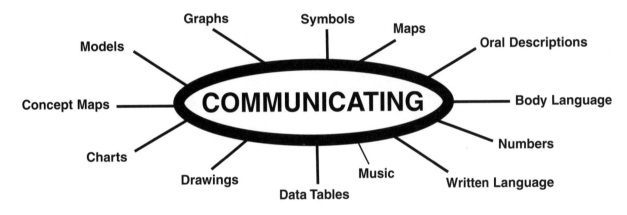

 Activity 2.4   Giving and Following Directions

In this activity, instead of just describing an object, you will practice giving directions to a partner. To begin, think of a procedure you might want someone to follow using your directions. Here are a few suggestions but you may want to use one not listed here. The procedure you decide to describe may be limited by what materials are available to you. See the first two examples.

B. C. by permission of Johnny Hart and Field Enterprises, Inc.

How to plant a seed. (You will need a cup, some soil, a seed, and water.)

How to tie a shoe. (You will need a shoe with laces.)

How to make a 2 liter bottle terrarium.

How to wash and dry your hands.

How to measure the length of your foot.

How to make a paper airplane.

How to make a simple electric circuit.

Do the following:

1. Select a partner to work with you.
2. Sit behind your partner so he or she cannot see you, or erect a screen or barrier between you and your partner.
3. Plan in your head how you will communicate the procedure to your partner.
4. Give precise directions for your partner to follow using whatever materials are appropriate. Your partner must do exactly what you say and *only* what you say.
5. When you are done, have your partner evaluate the quality of your procedure description using the 'Communicating Directions Rubric.' A check in the 'Yes' column means that criterion was met. A check in the 'Needs Improvement' column means you need to work on your communication skills.

Communicating Directions Rubric

Criteria	Yes	Needs Improvement
The Communicator . . . 1. Presents information in an organized way.		
2. Makes the description brief.		
3. Uses precise language.		
4. Communicates information accurately.		
5. Considers another's point of view.		

 Activity 2.5   Communicating with Maps

Some directions can be effectively communicated only with the use of maps. A map is any symbolic representation and includes mathematical formulas, patterns, guides, floor plans, blueprints, photographs, schematic drawing, and descriptions as well as more traditional land maps. In this activity you will use a number and letter grid to locate particular features on a traditional map. A map should have the following information:

1. a title, telling what the map is about
2. symbols, representing places or objects
3. a key, telling what each symbol represents
4. a scale, showing relative distances and sizes of objects
5. a direction rose, showing north, east, south and west

Mrs. Q, the teacher in the scenario at the beginning of this chapter, used several ideas from the Website listed below to help teach mapping skills to her students. She shared this Website as well as other sites and resources with parents and encouraged their use at home. Making maps helped the children to tell their stories about things they had investigated. One student used a map to describe the area under his bed. Toys and various other items, even dust bunnies, were appropriately identified in the map's key.

http://www.ed.gov/pubs/parents/Geography/index.html

Pictured here is a map similar to the one the class made when they visited a nearby pond. By placing numbers along the top of the map and letters along the side of the map, the children found it easier to tell their *key pals* where things were. The willows, for example, were at the north end of the pond at about B-5 and B-6. Under the willows students saw water flowing into the pond.

Deer Pond

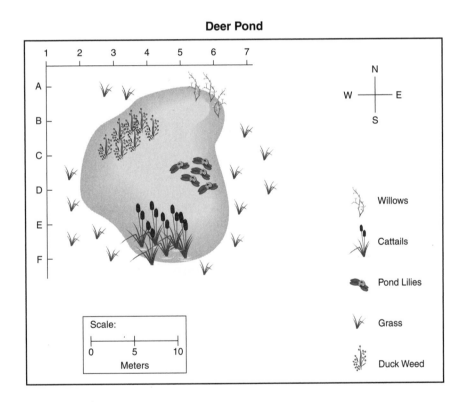

Use the map to find out more about what the class did at the pond. For each item that the children saw at Deer Pond, locate its position on the map and identify its location by naming the appropriate letter and number on the map.

1. _____ saw water flowing under the willows
2. _____ saw red winged blackbirds in the cattails
3. _____ saw a duck eating duckweed
4. _____ saw frogs in the pond lilies
5. _____ saw fish jumping in open water
6. _____ saw deer tracks along the southeast shore

Self-Check _____ Activity 2.5

1. B-5,-6
2. E-3,-4,-5
3. B-3,-4 and C-3,-4
4. C-5,-6 and D-5,-6
5. C-5 and D-3,-4
6. E-7

Activity 2.6   Practice Making and Following Maps

In this activity you will practice making a map for your partner to follow and you will practice following someone else's map. When you are done, your partner will evaluate your map.

1. Select a partner to work with you.
2. You and your partner should each place an object somewhere in the room but do not let each other see where it was put.
3. Make a map of the room showing the location of the object.
4. Exchange maps with your partner.
5. Use your partner's map to find his or her hidden object.
6. Once you and your partner have found the objects, evaluate each other's maps using the 'Communicating with Maps Rubric.' A check in the 'Yes' column means that criterion was met. A check in the 'Needs Improvement' column means you need to work on your mapping skills.

Communicating with Maps Rubric

Criteria	Yes	Needs Improvement
1. map has a title		
2. map has symbols representing places or objects		
3. map has a key telling what each symbol is		
4. map has a scale showing relative distances		
5. map has a direction rose showing north, east, south, west		

 # Self-Assessment

1. Select any object in the room and write a description of it.
2. Construct a map showing how you might exit the building in an emergency.

Self-Assessment Answers

Give your object description to someone and have them identify the object. Give your map to someone and have them use it to explain to someone else how to exit the building in an emergency.

IDEAS FOR YOUR CLASSROOM

1. Although it was not covered in this chapter, writing is a very important communication skill. Good clear writing like clear verbal communication, must be practiced. One way this can be practiced in science is to have students write what they are learning, especially during activities. You may want to adopt a format similar to the one below:

   a. We were studying _____
   b. We did _____
   c. We observed _____
   d. We learned _____

2. A game that will help students' descriptive skills is similar to one earlier in the chapter. Place two or three students in each group. Let one student pick out an object in the room and describe it to the others. (The description should be observations rather than function, that is, it is red rather than you write with it.) When the object is correctly identified, then another student gets a turn.

3. An interesting application of communication skills to social studies would be to obtain an old map of your geographic area and a current map. If you are lucky enough to get a map of your own area when your state adopted its seal (excepting Hawaii and Alaska), you get a bonus. Compare the maps. What changes have taken place? What used to be where you are now? Where is your home? Now for the bonus. . . Your state seal is a piece of communication. When it was adopted, the people of the state were trying to communicate the state's important qualities at that time. Have these qualities changed? If you were going to design a state seal, what would you include now?

4. Using our Resources
   a. *Where does it originate?* is an interesting question for a study. Bulletin boards and activities could be used to help students gain an appreciation and understanding of how we use our environment to meet our needs. Many children need to learn that stores are not the sources of eggs, milk, pencils, baseball bats, and clothes. In a study of the sources of foods, chains could be constructed to show events between the producers and consumers. (The chains could vary in level to fit the sophistication of the students.) If desired, steps could be added to illustrate processing, transportation, sales or any other event between the raw product and the consumer.

   Where does electricity originate? When you turn on a switch, does the electricity come from a local plant or one far away? What kind of plant generates the electricity? What kinds of energy transformations take place between its production and use? As you use electricity, what kinds of energy transformations take place?

b. *Where does it go?* is another question worth studying. What happens to things we "throw away"? This could lead to a study of waste disposal and some of its problems as well as to a study of recycling. While *What happens when I flush the toilet?* may not be one of the burning questions in your life, it might be worth exploring.

5. *Gossip* is an interesting game that helps improve communication skills. You could start the game by having the students form a circle. Then you give a short written message to one of the students. The student reads the message and whispers it once to his or her neighbor, who in turn passes the message along verbally. When the message has gone around the circle, the last student says the message aloud. Compare the original message with it. This could lead to a discussion of how we receive information. Perhaps it would be possible to visit a radio or television station or newspaper office. If that's not possible, maybe they have a speaker that could visit your class.

6. *Kidnap* is another game that can improve observation and communication skills. Have three students, one victim and two villains, perform the following skit. Dressed in special clothes, the masked villains kidnap the victim (also dressed in unique garb) quickly from the room. Have the class write a brief eyewitness account, describing the event and the descriptions of the victim and kidnappers. Compare the results. You might have a person from the police visit the class to talk about accuracy of communication and being an eyewitness.

7. Labels communicate! Have your students become label readers. Find out the contents of junk foods or any other food that comes in a container. What other products have labels? What does the label tell about the product?

8. Advertisements are another form of communication. Have your students study different advertisements from different sources: television, magazines, newspapers, etc. What are they communicating? Are there hidden messages?

Start Kids Thinking

What words would you use to describe this object so that someone else can identify it?

Describe everything you observed as completely as you can.

Using good descriptive words tell everything you observed about this object (or event).

What method would you use to communicate to someone else what you observed?

High Stakes Testing

A sample multiple-choice item from State Standardized Exams.

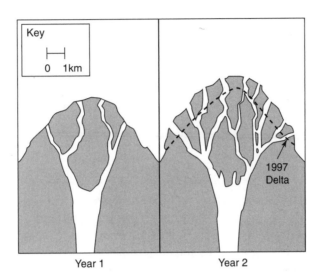

The picture shows the development of a delta over a two-year period. According to this information, about how far did the delta reach into the ocean after 1997?

    A   0.1 km

    B   0.5 km

    C   1 km ←

    D   1.5 km

Virginia: SOL Grade 5 Spring 2001 Release Item.

Technology Spotlight:

Enhancing Communication with a Digital Camera

A digital camera can do almost anything that a conventional film camera can do and more. By storing images in digital form they can be saved and edited to enhance learning, provide motivation, and promote creativity by students and teachers alike.

Digital images have numerous applications in the classroom. For example, students can document a change over time, such as a plant's growth from germination to seed production, or erosion that takes place in the schoolyard over several months. Another use is to capture the distinct steps of a science experiment and to put those images into a slide show to share with others. Through the use of digital images teachers can have students preview learning opportunities at upcoming field trips to set expectations and facilitate discussion, or to share experiences with others after the trip using their own images.

Although initially time consuming, teachers can incorporate pictures of objects and organisms into lessons to help students visualize the concepts being presented. Visual enhancement of traditional lecture-discussions reinforces learning for students with different learning styles. When images are incorporated in PowerPoint® lessons, absent students have visual cues to increase their understanding of the information they missed. Digital images are a convenient way of documenting a student or class project by importing images into PowerPoint® or Hyper Studio® for presentation during an open house, awards night, or science fair. With increasing emphasis on accountability, digital images make it easy to document individual student work for an assessment portfolio. Digital cameras permit the documentation of a greater variety of student work for a portfolio than would normally be possible with film cameras.

Digital cameras store images as jpeg files that can be saved on memory sticks, floppy disks, and CDs. Many cameras also provide for direct transfer of digital images to computers. The picture quality and storage capabilities of digital cameras are rapidly increasing as prices are decreasing. Use a search engine to locate current Internet sites that provide support and advice in purchasing as well as ideas for using digital cameras in the classroom.

http://www.plainfield.k12.in.us/hschool/webq/webq43/shannon.htm
www.uen.org/utahlink/lp_res/TRB035/html#communicating
md.essortment.com/communicationte_rqmd.htm
www.sasked.gov.sk.ca/docs/elemsci/grluaesc.html#actl
www.hhmi.org/coolscience/inchsquare/index.html
www.webofroses.com/scouting/communicating.html

WEBSITES FOR ACTIVITIES

A Model for Assessing Student Learning

Assessment Type: Pencil & Paper Teacher Observation Checklist

Observational checklists may be used to assess students' acquisition and use of specific skills. Use a check mark or date notation to document appropriate process skill behaviors as you observe them. Compare this paper and pencil approach to assessment with a new technological method for recording teacher observations that is illustrated in Chapter 11.

Student Name	Uses appropriate vocabulary to describe objects and events	Communicates clearly	Listens to others	Selects and uses appropriate communication tools

Chapter 3

Classifying

National and State Standards Connections

- Objects can be described by the properties of the materials from which they are made, and those properties can be used to separate or sort a group of objects or materials. (NSES K–4)
- Sort, classify, and order objects by size, number, and other properties. (NCTM Pre-K–2)
- Plan and conduct investigations in which objects are classified and arranged according to attributes or properties. (Virginia Standards of Learning, Grade 1)

MATERIALS NEEDED

For the activities in this chapter you will need the following:

- ✔ a set of 6 assorted shells (numbered 1,2,3,4,5,6)
- ✔ a set of information panels from cereal boxes
- ✔ 6 peanuts in the shell
- ✔ 4 or 5 people to observe
- ✔ an assortment of pasta shapes

Questions Kids Ask

"Is that green thing in the water a plant or an animal?"

"Is a daddy-long legs a spider?"

"If I got lost in the woods, which plants could I eat?"

"I caught a perch. What kind of fish did you catch?"

"What is the difference between a toad and a frog?"

"Besides water, what else is a liquid?"

"Which magnet is the strongest?"

The human brain is a sorting tool. It is able to take information sent to it by the senses, sort it out, and make sense of it. New information for which the brain has already created a category or grouping, gets placed in that category. Information for which the brain has not already created a category may get lost. But the brain can create new categories. The brain forms these categories to prevent this loss by recognizing how things are similar to one another, how they are different from one another, or how they relate to one another. The more categories the brain has, the more sense it can make of 'new stuff.'

An effective teacher helps children learn *how* to set up useful mental categories, ones that not only make sense of immediate experiences but also serve to accommodate new information. A child who encounters a long, squirmy, skinny, brown, slimy, segmented thing recognizes it as a type of 'worm,' a category his or her brain has set up to accommodate long, squirmy, skinny, brown, slimy, segmented things. A child who thinks it is a snake has not yet learned the attributes, or characteristics, that distinguish a worm from a snake. Conveniently, a child's brain can set up a category for 'worm' based on a certain set of characteristics and then be able to recognize an organism as a worm whenever it squiggles by. The child does not have to relearn 'worm' every time one is encountered. Classification is essentially the basis for all concept formation.

In this chapter we will examine three basic forms of classification. The first, **Single Stage Classification,** involves separating a set of objects into two or more subsets on the basis of at least one observable property. For example, seeds are separated into two groups as either one-part or two-part seeds, monocotyledons and dicotyledons, while clouds are grouped into three basic groups as cumulus, stratus, and cirrus clouds. The *simplest* form of single stage classification is *binary* classification where a group of objects is sorted into two groups on the basis of whether each object has or does not have a particular property; binary is a yes/no category system. Two examples are animals with and without backbones and buttons that have two holes and those that do not. Binary classification is a special kind of Single Stage classification.

Multistage Classification occurs when sets are sorted into subsets and then each subset is sorted again and again, creating several layers or stages of subsets. For example, buttons already sorted by number of holes may then be separated into those that are plastic and those that are not. In science animals that have been grouped as vertebrates might be further separated into groups of fish, amphibians, reptiles, birds, and mammals. All of which are subsets of vertebrates.

Serial Ordering is a kind of classification where objects are placed in order by the extent to which they possess a particular property, such as diameter or mass. Buttons or rocks, for example, may be placed in order from smallest to largest or from heaviest to lightest.

B. C. by permission of Johnny Hart and Field Enterprises. Inc.

Goal

The purpose of these exercises is to help you learn to classify objects and events on the basis of observable characteristics.

Performance Objectives

After completing this set of activities, you should be able to:

1. Given a set of objects, list observable properties that could be used to classify the objects.
2. Given a set of objects, construct binary as well as other single stage classification schemes for the objects.
3. Given a set of objects, construct a multistage classification scheme and identify the properties on which the classification is based.
4. Given a set of objects, identify properties by which the set of objects could be serially ordered and construct a serial order for each property.

Activity 3.1 Constructing a Binary Classification System Based on Observable Properties

Single stage classification of a set of objects results in the formation of two or more sub-groups. The simplest form of this type of classification is a yes/no category system called *binary classification.*

In a binary classification system the set of objects is divided into two subsets on the basis of whether each object has or does not have a particular property. To construct a binary classification system you must first identify a common property that only some of the objects have. Then group all the objects displaying that property in one set and all the objects not displaying that property in another set. For example, biologists classify living things into two groups: animals and plants (plants being the group *not* displaying animal properties). Scientists further classify animals into two groups: those with backbones and those without backbones. When constructing a binary classification system, be certain that all the objects in the original set will fit into one and only one of the two subsets. This is shown in the following activities.

➡️**GO TO** the supply area and obtain a set of shells. If you do not have real shells, use those pictured with the activity.

Look at the set of shells on the following page and observe their similarities and differences. In the left-hand column of the chart provided, list at least three observable properties by which the shells can be grouped into two subsets. In the *Yes* column write the numbers of the shells that have the property that you have identified. In the No column, write the numbers of the shells that lack the property you have identified. Examine the following example. Binary classification is essentially a yes/no grouping system. Yes, they have the properties; no, they do not have the properties.

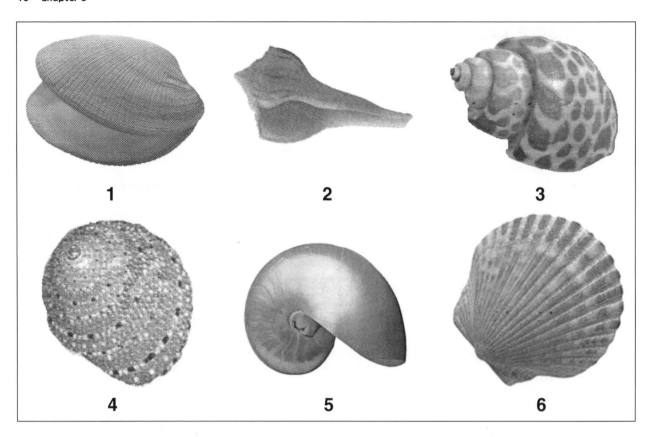

Observable Properties	Yes	No
1. has dark spots	3, 4	1, 2, 5, 6

Notice, as in the example, that for each property the subsets accommodate all the objects in the original set and that every object can be assigned to one of the subsets.

Compare your answers with someone else's. If you used the shells pictured, check your observations with those that follow.

Self-Check _____ Activity 3.1

Some of the properties you may have identified are as follows. Be certain that for each property the subsets accommodate all the objects in the original set, and that every object can be assigned to one and only one subset.

Observable Properties	Yes	No
1. has dark spots	3, 4,	1, 2, 5, 6
2. spiral shaped	2, 3, 4, 5	1, 6
3. surface is ribbed	1, 6	1, 3, 4, 5
4. scalloped edges	6	1, 2, 3, 4, 5

Mapping an Organization Scheme

Charts, tables, and diagrams are very useful communication tools. Notice that the previous chart is binary in nature because it sorts items into either one category or another according to whether or not each item possesses a certain characteristic.

Another powerful communicator is a graphic organizer or map. Notice that the map diagrammed below is also binary because the original set is split into only two subsets. The map clearly identifies which properties are being used to sort the items and indicates which items belong in each subset.

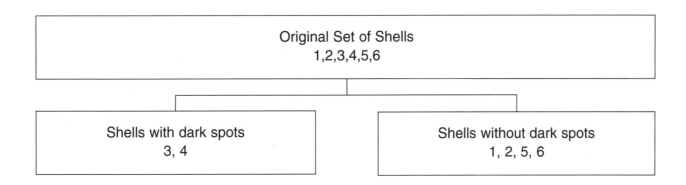

Original Set of Shells
1,2,3,4,5,6

Shells with dark spots
3, 4

Shells without dark spots
1, 2, 5, 6

Activity 3.2 Using Several Properties at Once

Binary classification can also be used when you wish to group objects that have more than one property or characteristic in common. For this activity, instead of using objects, you will use people. Use yourself and four other people seated near you. When talking about people we usually refer to properties as characteristics. Those people having several characteristics in common may be sorted into one group and those people not having those characteristics may be sorted into the other group. The first group, for example, may contain all the people who have brown hair *and* blue eyes. The second group will contain all the people from the original set who do not have *both* of these characteristics.

1. Using yourself and four other people, what are some ways the people can be grouped using two or more characteristics?

   _____

   _____

   _____

   _____

   _____

2. Construct a map to show a binary classification based on several shared characteristics of the people. The box showing the original set has been drawn for you but you will need to write names or initials. Then complete the scheme making sure it is binary and that each subset contains the specific characteristics and the names of the people possessing those characteristics.

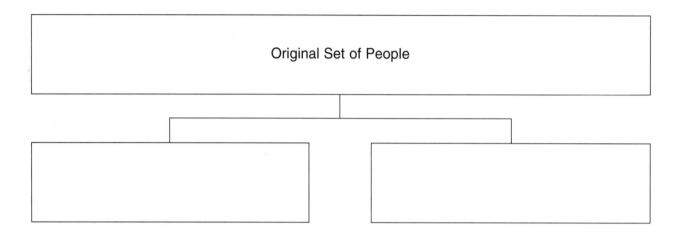

Self-Check _____ Activity 3.2

Answers will vary. They could include:

1. brown eyes and brown hair
2. over 150 cm (~5 feet) tall and female
3. blonde hair and fair skin
4. the answers could also be less obvious. For example, one set could be round face, wears glasses, and long eyelashes.

Your map will differ according to which people you used and the characteristics you chose, but here is an example.

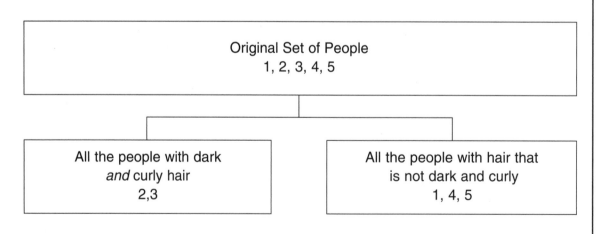

Another way to check your map is to give it to someone else. If they are able to use the map to match the characteristics with the people, it must be a good one.

Activity 3.3 Constructing a Multi-Stage Classification System

Performing a binary classification or any other type of single stage classification on a set of objects and then again on each of the subsets results in a classification scheme consisting of layers or stages. When objects are classified in a single stage way again and again, a hierarchy of sets and subsets is created, called a *multi-stage classification*. When a binary type of single stage classification is used, subsets are determined by sorting objects that have a particular property from those that do not have that property.

Animals, for example, are classified as either having backbones or not having backbones. Continuing with binary classification could result in those animals having backbones being further classified as either having or not having hair. But remember, not all single stage classification has to be binary. Materials that are attracted to a magnet could be classified as being made of iron, steel, or nickel. Those materials made of iron could be further classified as those that occur naturally and those that were manufactured.

Because you are already familiar with the shells in Activity 3.1, we will use them to illustrate how a multi-stage classification scheme is constructed. Notice that the scheme identifies the observable properties by which the items are sorted and that each item associated with the properties is identified.

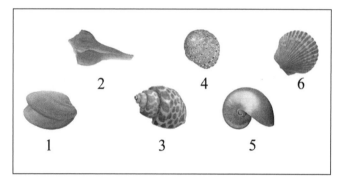

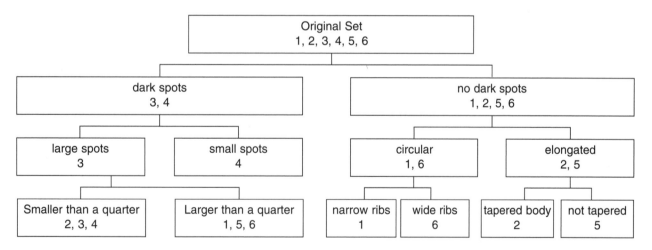

In addition to the characteristics already discussed, a multi-stage classification system has the following features:

1. Several other schemes may be possible depending upon which observable properties are used for grouping.
2. When each object in the original set is separated into a category by itself, the scheme is complete.
3. A unique description of each object can be obtained by listing all the properties that the object has. In the previous scheme, for example, shell # 6 can be distinguished from the other shells in the original set by listing its properties: no dark spots, circular, wide ribs.

Now it is your turn to construct a multi-stage scheme. You can use the same people you used in Activity 3.2 or you can use different people. Complete the classification scheme started below. In each box, indicate the characteristic, such as blue eyes, that you used to make the grouping. Be sure to carry the scheme through to completion. *You may need to add more boxes or delete some.* The number of boxes you will need depends on whether you are using a binary classification scheme or another form of single stage classification in which each set is divided into two *or more* sub-sets based upon the characteristics they possess. When the scheme is completed, each person should be in a separate box.

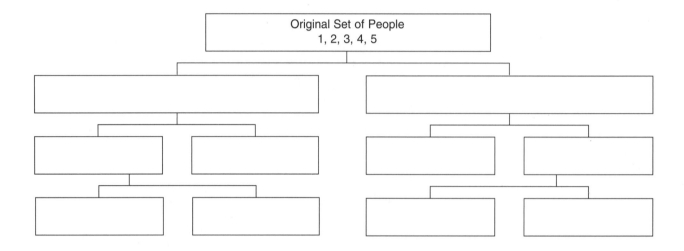

Original Set of People
1, 2, 3, 4, 5

Self-Check _____ Activity 3.3

There are many possible schemes depending on the person's characteristics and the order in which you selected them. If in doubt, ask your instructor.

For additional practice, do the classification activity below. You will need sharp observation skills!

Activity 3.4 Multi-Stage Classification Additional Practice

→**GO TO** the supply area and obtain a set of 6 peanuts in the shell.
Observe your set of peanuts carefully.

Construct a multi-stage classification scheme for your peanuts making sure your scheme fulfills these requirements:

- The items in the original set are identified.
- Each sorting action is binary. (yes/no) or other form of single stage classification (e.g., 1, 2, or 3 bumps)
- Each subset is identified by one or more properties and the items possessing those properties.
- The scheme is continued until each item is in its own subset.

If your set of peanuts is not pre-numbered, you should number them 1–6.

```
┌──────────────────────────────────────────────┐
│              Original Set of Peanuts           │
│                      1–6                       │
└──────────────────────────────────────────────┘
```

Self-Check _____ Activity 3.4

The best way to check your map is to give it and your set of peanuts to someone else and have them use it to classify the peanuts.

✔_____

Activity 3.5   Serial Ordering

It sometimes is preferable to order objects according to the extent to which they display a particular property. Depending upon the purpose of the classification, objects may be ordered on the basis of size, shape, color, or a variety of other characteristics. In a hardware store, nails are ordered on the basis of size. Paints can be arranged according to size of can or color. Clothing stores use size to arrange merchandise in serial order. In science, minerals are ordered by degree of hardness (Moh's hardness scale) and metals can be ordered by amount of heat conductivity.

➡ **GO TO** the supply area and obtain a set of information panels from cereal boxes. Examine them and identify three properties by which these panels could be arranged in order.

1. _____

2. _____

3. _____

Self-Check _____ Activity 3.5

Answers will vary. They could include:

1. number of calories per serving.
2. amount of iron per serving.
3. amount of thiamin per serving.
4. amount of protein per serving.
5. the length of the panel itself.

✔_____

One way to graphically show a serial ordering is with an arrow that is labeled to indicate an increase or decrease in the extent to which items display a certain property. You might, for example, serial order students in your class according to height. That serial ordering might look like this:

Bob   Leticia   Joe   Bill   Amy   Maria   Ashley   Lamar

Shortest ➡ **Tallest**

Height

Serial order the panels of cereal boxes according to one of the properties you identified earlier. Represent the serial order on the arrow shown. Be sure to label the *property* that all the items display, the *degree* to which they display the property, and each item's *position of rank* along the continuum.

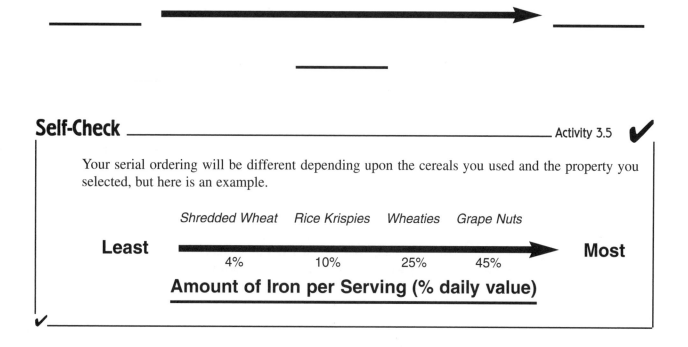

Self-Check ——————————————————— Activity 3.5 ✔

Your serial ordering will be different depending upon the cereals you used and the property you selected, but here is an example.

Shredded Wheat   *Rice Krispies*   *Wheaties*   *Grape Nuts*

Least   4%   10%   25%   45%   **Most**

Amount of Iron per Serving (% daily value)

Self-Assessment

Classifying

For this test you will construct: (a) a binary classification scheme; (b) a multi-stage classification scheme; and (c) a serial order classification scheme for a set of pasta shapes.

➡**GO TO** the supply area and obtain a set of pasta shapes (shell, spiral, elbow, wheel, tube, bowtie).

a. In the chart below identify at least three observable properties by which these shapes could be classified in a binary classification scheme. In the proper column indicate which shapes have or do not have each property.

Observable Properties	Yes	No
1.		
2.		
3.		

b. On a separate piece of paper or below, construct at least one multi-stage classification scheme for the pasta shapes and carry it through to completion. In each box identify the property used for grouping and list the name of each shape with that property.

c. Identify a property by which the pasta pieces can be serial ordered. Then serial order the pieces on the basis of that property. In the space below draw and label an arrow that accurately communicates how you serial ordered your pasta pieces.

Self-Assessment Answers

a. Some observable properties on which binary classification schemes for these pieces of pasta could be based are listed below (our pasta may differ slightly from your pasta):

Observable Properties	Yes	No
cylindrical shape	tube, wheel, elbow	spiral, shell, bowtie
twists or turns	tube, elbow, spiral	shell, wheel, bowtie
has compartments	wheel	spiral, shell, elbow, bowtie, tube
ribbed surfaces	tube, wheel, shell	spiral, elbow, bowtie

b. One possible multi-stage classification scheme for a set of pasta pieces is shown below. Many other schemes are possible depending on the pasta products chosen and the properties used for grouping. Our multi-stage scheme is a series of binary classifications that resulted in two sub sets each time, but ours or yours could easily be grouped into three or more subsets at each level. If you have questions about your classification scheme, compare it with someone else's, or see your instructor.

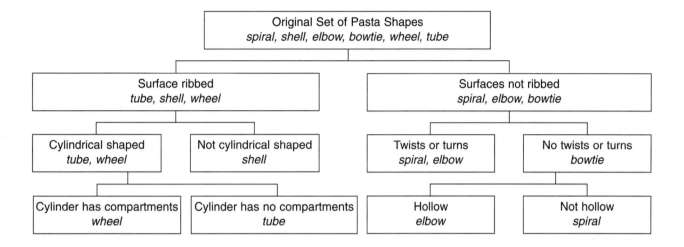

c. A couple of ways of serial ordering your pasta pieces are shown below (again our pasta may differ from yours):

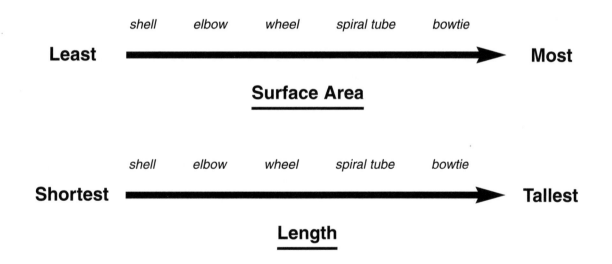

IDEAS FOR YOUR CLASSROOM

1. Water play is an enjoyable activity for young learners. In one set of activities, students can compare objects that sink with objects that float. (This is an early experience that later leads to concepts of density and buoyancy.)

   In another set of activities, students can predict the serial order of the volumes of empty containers and check their prediction by filling the containers, one with the others. (In doing this, they learn how to relate sizes and shapes with volumes of containers.)

2. Liquids vary in density and this affects their buoyant properties. An intriguing puzzle for older children is to present them with two containers of water: one in which a hard boiled egg is floating and the other in which the hard boiled egg is resting on the bottom. (The only difference between the glasses of water is that one contains a few tablespoons of dissolved salt.)

3. Animals are grouped on the basis of similarities and differences. Ways in which they may be grouped include: how they obtain their food; whether they lay eggs or bear live young; by their habitat (parks, forests, oceans, ponds, and so on); or how they protect themselves. (Have you ever wondered why a zebra has stripes?)

   Collect pictures of various animals and explore the various ways they can be classified using single stage classifications. If your class feels really ambitious, they may construct a multi-stage classification system for animals. (You may want to limit their efforts to local kinds of animals because the entire animal kingdom is a gigantic multi-stage classification system.)

4. Plants can be used for single stage, multi-stage classification and serial ordering. Plant leaves are excellent for single stage and multi-stage classification. Trees, for example, could be serial ordered by the distance around, height, age (by counting the rings) and a number of other ways.

   One interesting project would be to investigate how plants disperse their seeds: some rely on the wind; some need animals to eat their fruit; some "hitch" rides on animals; just to mention a few. Further, examine the similarities and differences of plant seeds dispersed in the same manner. Maple seeds and dandelions are both spread by the wind but their methods are different. Dandelion seeds could be likened to parachutes and maple seeds could be compared to helicopters.

5. Is it natural or human-made? Some fabrics from which clothes are made occur naturally while others are manufactured from other products like petroleum.

6. Nutrition is a natural unit to practice classification skills. Besides studying the food groups, you can have your students read the sugar content of various cereals and measure out equivalent amounts of sugar. (Sand or salt may be substituted to give the same visual effect at a lower cost.) Once the amounts have been measured out, they can be ordered.

7. Have your students classify rubber "bugs," available in toy stores or nature shops.

- How is this object like that one?
- How are these objects alike; how are they different?
- Sort these items according to their properties.
- Put these objects in order and explain the order.
- Group these objects by color (size, texture, smell, taste, or sound they make).
- Classify this set of objects in as many different ways as you can.
- How would you construct a multi-stage classification scheme for this collection of objects?

High Stakes Testing

A sample multiple-choice item from State Standardized Exams.

Classification Key

1a Body kite-like in shape	Ray
1b Body not kite-like in shape	Go to 2
2a Nose saw-like in shape	Swordfish
2b Nose not saw-like in shape	Go to 3
3a Head extended on both sides	Hammerhead shark
3b Head not extended on both sides	Go to 4
4a Body has spots	Leopard shark
4b Body does not have spots	Nurse shark

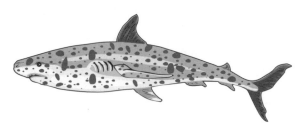

Using the picture and classification key, what is this animal?

A   Swordfish
B   Hammerhead shark
C   Leopard shark
D   Nurse shark

Virginia: SOL Grade 5 Spring 2001 Release Item.

Technology Spotlight

Using Software to Practice Classification Skills

Kidspiration™ is a visual learning tool especially for K–3 students. Designed for emerging readers and writers, Kidspiration™ helps students to organize information, understand concepts and connections, create stories, and express and share their thoughts. Integrating pictures and writing enhance students' comprehension of concepts and information. Kidspiration™ can be used to build students' skills in the core areas of reading and writing, science, and social studies. The version for older students and teachers is called Inspiration™.

If a copy of this software is available, load Kidspiration™ (or Inspiration™) software in a computer and spend some time perusing its many uses. If it is not available, go to *www.inspiration.com* and click on Kidspiration on the left-hand side. After you have explored the site, download a trial version from the site. Open the program and spend some time exploring the functions of the various buttons. See if you can create a multi-stage classification scheme on some topic using this software. Here is one we did on rocks:

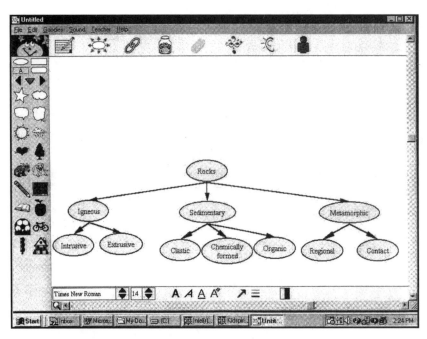

http://library.thinkquest.org/J002289/index.html
http://www.hhmi.org/coolscience/critters/index.html
www.designastudy.com/teaching/tips-1000.html
www.fi.edu/tfi/units/life/classify/clasact2.html
www.successlink.org/great2/g1290.html
www.monroe2boces.org/shared/esp/beansort.htm

WEBSITES FOR ACTIVITIES

A Model for Assessing Student Learning

Assessment Type: Performance Task

Teacher Preparation

The amount of materials needed will depend on how many students are doing the task at the same time. The most economical strategy is to use stations through which students can rotate. Re-sealable clear plastic bags, paper or plastic plates, and a variety of beans are needed. When selecting beans, choose beans of different shapes, colors, and sizes.

Student Directions

1. Check your materials.

    ■ Bag of beans

    ■ Paper or plastic plate

2. Empty the bag of materials onto the plate.

3. Observe the beans. Put the beans into two groups so that something is the same about all the beans in each group.

    ■ Write one way all the beans you put into group 1 are the same.

    _____

    ■ Write one way all the beans you put into group 2 are the same.

    _____

4. Put the beans all back together on the plate. Observe them again. Think of another way to put the beans into two groups so that there is something the same about all the beans in each group. Put the beans into two groups.

    ■ Write another way all the beans you put in group 1 are the same.

    _____

    ■ Write another way all the beans you put in group 2 are the same.

    _____

5. When you are done, put the beans back into the bag.

Scoring Procedure

For #3   Student identifies a property that all beans in group 1 have in common. Sample acceptable answers are: small, large, long, round, dark, brown, not green, broken, whole, and so on. For group 2, students use any property that they did not use for group 1.

For #4   On both items, students use a property (see sample answers above) not previously used.

Chapter 4

Measuring Metrically

National and State Standards Connections

- Employ simple equipment and tools to gather data and extend the senses. (NSES K–4)
- Understand measurable attributes of objects and the units, systems, and processes of measurement. (NCTM 3–5)
- Measure length, weight, temperature and liquid volume with appropriate tools and express those measurements in standard metric system units. (California Content Standards, Grade 2)

MATERIALS NEEDED

- ✔ a meter stick
- ✔ a metric ruler
- ✔ an equal-arm balance
- ✔ a set of masses
- ✔ 20 centicubes
- ✔ a liter container
- ✔ 4 containers of various sizes and shapes
- ✔ a graduated cylinder
- ✔ ice cubes

- ✔ a Celsius thermometer
- ✔ 2 small rocks; one solid, one porous
- ✔ a can of soda
- ✔ a bag of potato chips
- ✔ a container of cookies
- ✔ a container of sugar
- ✔ a container of vegetable shortening (solid at room temperature)
- ✔ a spring scale calibrated in newtons

Optional, to make an overflow container: a 2 Liter bottle, bendable straw, a standard paper hole punch, scissors

For the self-assessment you will need a *Measurement Test* bag containing:

- ✔ Ziplock sealable sandwich bags
- ✔ a small plastic disposable cup

- ✔ supply of gravel, enough to fill a small plastic cup

B. C. by permission of Johnny Hart and Field Enterprises, Inc.

Questions Kids Ask

Does water take up more room when it freezes into an ice cube?

Is my right hand bigger than my left hand?

Does a cooked egg have more stuff (mass) in it than a raw egg?

If I weigh myself and then eat 2 pounds of lunch, will I weigh 2 pounds more than I did before lunch?

How much trash does our school throw out in a day?

In a 10K race, what's a 'K'?

The children who asked these questions can answer their own questions when they are taught to measure accurately and to use appropriate measuring units. Encouraging children to ask and to answer their own questions helps them to become better learners.

The metric system serves as the language of measurement for most of the world. The metric system gives us easy to learn units for everyday use. Multiplying and dividing are relatively easy operations because the metric system is in base ten. In order to help your students become comfortable using the metric system, you must first become comfortable with it yourself.

Goals

In these exercises you will learn and practice skills needed to do measurements in the metric system. As you develop these skills you should begin to think metric.

Performance Objectives

After completing this set of activities, you should be able to:

1. Select the appropriate metric unit for measuring any property (length, volume, temperature, mass, and weight) of a given object.
2. Given a set of metric units, state equivalent metric measures using prefixes (that is, perform conversions *within* the metric system).
3. Measure the temperature, length, volume, mass, or force of any object to the nearest tenth (0.1) unit.

Activity 4.1   A Metric Treasure Hunt

While you are studying this chapter and learning more about the metric system, find out just how metric your world already is. Every chance you get look closely at labels on products in your home, at school, and in stores where you shop. As you find products that fit the following descriptions, name the product and record the unit measurement part of the label.

Find a product whose label has the English measurement written first and the metric measurement written in parentheses; for example, the label on a sack of flour may read: 5 lbs. (2.26 kg). _____

_____

Find a product whose label has the metric measurement written first and the English measurement written in parentheses; for example, the label on a shampoo bottle may read: 946 mL (32 fl. oz.). _____

_Bottle of water 500ml (16.9 FL OZ)_____

Find a product whose label shows that its contents are in a whole metric quantity; for example, the label on a bottle of water may read: 1 Liter (33.8 fl. oz.). _____

_____

Find a product whose label shows weight in customary (English) units, but then shows mass instead of weight in metric units; for example, 13 oz. (370 grams). _____

_____

Find a product that is labeled with only the metric measurement. _____

_____

Find a product that is labeled with only the customary (English) measurement. _____

_____

Activity 4.2   Measuring Metrically

The modern version of the metric system is called Systeme Internationale d' Unites (international system of units), symbolized as SI. The term metric comes from the base unit of length in the system, the *meter*. The meter was originally defined as one ten-millionth of the distance from the equator to the north pole along a meridian that passes through France.

Two spellings of meter (and also liter) are recognized in the United States—meter or metre and liter or litre. Internationally, metre and litre are preferred, but in the United States the spellings of these units are most commonly meter and liter. Although most elementary science and mathematics textbooks and activity books use the spelling meter and liter, you should be comfortable with both spellings and acquaint your students with these differences.

SI uses *base units* to designate quantities being measured:

- The base unit for measuring length is the meter (symbol m).
- The base unit for measuring volume is the liter (symbol L).
- The base unit for measuring mass is the kilogram (symbol kg).

(Note: The reason the base unit of mass is not the gram is because the gram is a very small amount of mass, such as the mass of a paperclip.)

Though not discussed here, there are other base units in the metric system, including those used for measuring electric current, light intensity and temperature. And there are yet other units, called *derived units,* like the unit used for measuring force, that are made by combining base units. Let's keep it simple for now.

Next to each measurement shown below write whether it should be measured in meters, liters, or kilograms.

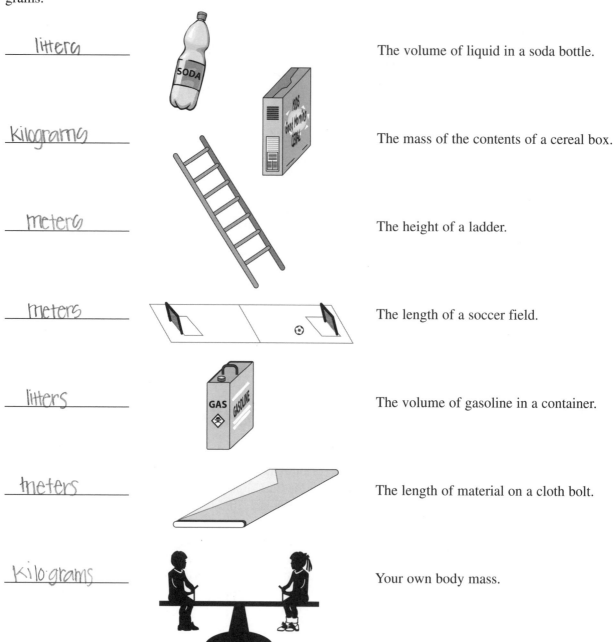

litera   The volume of liquid in a soda bottle.

kilograms   The mass of the contents of a cereal box.

meters   The height of a ladder.

meters   The length of a soccer field.

litters   The volume of gasoline in a container.

meters   The length of material on a cloth bolt.

kilograms   Your own body mass.

Self-Check ————————————————————————— Activity 4.2

soda: liter
cereal: kilograms
ladder: meters
soccer field: meters

gasoline: liters
cloth: meters
body: kilograms

 **Activity 4.3 Metric Prefixes**

When measuring something small in size, it is convenient to use a unit of measure that is small. When measuring something large in size, it is convenient to use a unit of measure that is large. At times it is even necessary to convert smaller units to larger units or larger units to smaller units. One of the advantages of the metric system is that its terminology makes conversions easy. Instead of having to remember conversions, like 12 inches equals 1 foot equals 1/3 yards equals 1/5280 mile, the metric system uses prefixes to convert to larger and smaller units of measure. By acting as a multiplier, a prefix changes the meaning of the base unit, making it either larger or smaller by powers of ten. Learning to make conversions within the metric system is just a matter of learning the names of the prefixes and knowing how each prefix affects its base unit.

The prefixes between 1/1000 of a unit to 1,000 times a unit are:

Prefix name	Prefix symbol		Prefix meaning
kilo	k	_____	1000 times the base unit
hecto	h	_____	100 times the base unit
deka	da	_____	10 times the base unit

(no prefix) . . . measurement unit (meter, liter, gram[1]) . . . 1 times the base unit

deci	d	_____	1/10 of the base unit
centi	c	_____	1/100 of the base unit
milli	m	_____	1/1000 of the base unit

As you can see . . . kilo
                  hecto   . . . make the base unit larger
                  deka

                  deci
                  centi   . . . make the base unit smaller
                  milli

[1]There is one exception to the rule for adding prefixes to base units. Recall that the base unit for measuring mass is the kilogram, not the gram. The kilogram is the only base unit with a prefix already part of its name. When measuring mass, the prefix name is added to *gram* to avoid the use of multiple prefixes. Prefix symbols are used with the symbol *g* for gram, such as kg for kilogram, mg for milligram. With this exception, any of the prefixes can be used with any base unit.

With the advent of the computer age, other prefixes are becoming more frequently used; units such as *giga, mega, micro, nano,* which mean a billion, a million, one millionth, and one billionth, respectively. You may wish to place these prefixes in the previous chart.

Note from the chart that each prefix has a symbol that can be used in combination with the base unit symbol, kL for kiloliter and cm for centimeter for example.

Here is an example of a prefix making a base unit larger and more convenient to use for a particular purpose. Add the prefix *kilo* to the base unit *meter* and the result is *kilometer,* a unit 1000 times the length of a meter. Rather than saying Boston is 350,000 meters from New York, then, we can say that distance is 350 kilometers, or 350 km.

Here is an example of a prefix making a base unit smaller and more convenient to use for a particular purpose. Add the prefix *milli* to the base unit *meter* and the result is *millimeter,* a unit 1/1000th the length of a meter. Rather than saying a compact disk (CD) is about 1/1000th of a meter thick, we can say its thickness is about a millimeter, or 1 mm.

Practice saying prefixes with base units and explain to someone else what each one means.

The metric system is much like our monetary system because both are decimal (based on ten) systems. For example:

> Just as . . . 1 dollar = 10 dimes or 100 cents,
> so . . .   1 meter = 10 decimeters or 100 centimeters

You will be using the prefixes with each of the metric base units so spend a few minutes right now getting to know them and their meanings. Then do the practice exercise that follows.

Use your knowledge of the prefixes and their meanings to answer the following:

1. One meaning of *decimate* is to reduce something to what part of its original size?

2. A *decapod* is an animal that has how many legs?

3. In the decimal system, how is 0.3 read?

4. A *hectometer* equals how many meters?

5. A *kilowatt* equals how many watts?

6. If a *centipede* really had as many legs as its name implies, each leg is what part of the total number of legs?

7. On the *Celsius* temperature scale (formerly called *centigrade*), what part of the whole scale is one degree?

8. In a *millennium,* what part of the total time period is one year?

9. A *mill* is what part of a dollar?

Check your answers with the ones in the Self-Check.

Self-Check

1. decimate: 1/10 (deci) the original size
2. decapod: 10 (deka or deca) legs
3. decimal: tenths (deci)
4. hectometer: 100 (hecto) meters
5. kilowatt: 1000 (kilo) watts
6. centipede: 1/100 (centi)
7. centigrade: 1/100 (centi)
8. millennium: 1/1000 (milli)
9. mill: 1/1000 (milli) of a dollar

✔

Activity 4.4 Measuring Lengths Metrically

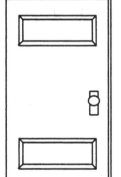

The meter is the basic unit for measuring length in the metric system. A meter is about the distance from the floor to a doorknob.

The symbol *m* is used for meter.

➡ **GO TO** the supply area and pick up a meter stick. Carefully observe the length of the meter stick, then close your eyes and try to picture that length in your mind. Look around the room and try to select some things which you think are about one meter in length. Then use the meter stick to check your estimations. When you are through, you should have a good picture in mind of how long a meter really is. If you want to learn to *think metric,* always estimate the size of objects before you actually measure them. If your estimates are too large or too small when compared to the actual measurements, you have the opportunity to adjust your thinking about the size of things in metric units. In time, the habit of estimating first and then measuring will give you the ability to *think metric.*

Estimate the following lengths to the nearest meter and record them in the column labeled *estimate.* Then *measure* each of the lengths to the nearest meter and record the measurements in the ***measure*** column.

	Estimate	**Measure**
Length of instructor's table	6 ft	6 ft 16"
Width of the instructor's table	1.5 ft	2 ft.
Width of the doorway	3 ft	3'
Height of the doorway	8'	8'4"
Distance from floor to window sill	4'	3'

If your estimates and measurements are fairly close, you are beginning to think metric!!

Compare your measurements with someone else's.

You may already have noticed that it may be difficult to measure distances much longer or shorter than one meter using a meter stick. Distances longer or shorter than one meter can be described by using the metric prefixes. Distances much shorter than a meter can be measured using a metric ruler.

Now examine the meter stick or a metric ruler, and find the millimeter, centimeter, and decimeter marks on it. Depending on the manufacturer, you may or may not see unit names or symbols, such as millimeter or mm.

The distance between the *millimeter* marks is about as wide as the wire in a paper clip. It is about the distance between two legs of the letter *m*. It takes 10 millimeters to make a centimeter. Millimeter is symbolized *mm*. Although the English system of measurement uses abbreviations with periods (in., yd., oz., lb.), the metric system uses symbols without periods. The only exception occurs when a symbol is used at the end of a sentence.

The *centimeter* marks are about the same length as the width of a paper clip or the width of your little finger. It takes 100 centimeters to make a meter. Centimeter is symbolized *cm*.

The *decimeter* is a little longer than the width of your hand or about the length of a new piece of chalk. It takes 10 decimeters to make a meter. Decimeter is symbolized *dm*.

Use the following scales to compare these lengths. Again, try to picture in your mind how long millimeters, centimeters, and decimeters really are.

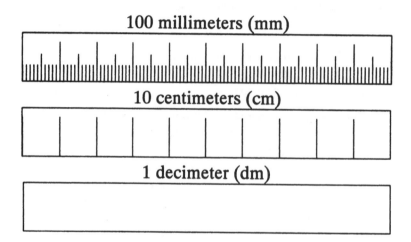

Now see if you can put what you have learned to use. Suppose you are asked to measure this line:

You should carefully lay your metric ruler along the line like this:

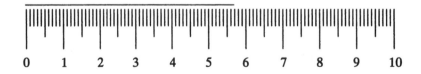

Notice that the line measures more than 5 but less than 6 centimeters. If you were to measure this line to the nearest centimeter, you would say it is 6 cm long because it is closer to 6 cm than 5 cm. More precisely the line measures seven millimeter marks beyond 5 cm. Rather than saying 5 cm and 7 mm, you should say 5.7 centimeters, or 57 mm. It is considered bad form to mix metric units. When measuring you will have to decide, or be told, how precisely you should measure, for example, to the nearest cm or to the nearest mm.

For practice. . .

On the scale below, identify and label 1 mm, 1 cm, and 1 dm. Then complete the following metric equivalents.

0   1   2   3   4   5   6   7   8   9   10

1. 1 meter = ____ decimeter = ____ centimeters = ____ millimeters.

2. 1 dm = ____ cm = ____ mm.

3. 1 mm= ____ m = ____ dm = ____ cm.

4. This page measures ____ centimeters wide. (Measure to the nearest cm.)

5. To the nearest 0.1 of a centimeter, this page is ____ cm long.

Check your answers with the ones following.

Self-Check _____ Activity 4.4

1. 1 m = 10 dm = 100 cm = 1000 mm
2. 1 dm = 10 cm = 100 mm
3. 1 mm = 0.001m = 0.01dm = 0.1cm or (1/1000 m = 1/100 dm = 1/10 cm)
4. 21 cm
5. 27.7 cm

What about measuring longer distances?

Distances longer than a few meters can be measured using a long metric tape or a trundle wheel calibrated in meters, centimeters, and millimeters.

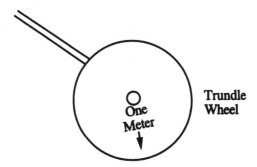

One Meter

Trundle Wheel

Long distances of several meters may be measured in dekameters or hectometers, although these units are not frequently used.

The *dekameter* is a measure of length equal to 10 meters. Its symbol is *dam.*

The *hectometer* is a measure of length equal to 100 meters. Its symbol is *hm.*

A race covering a distance of 500 meters is more likely to be called a 500 meter race than a 50 dekameter race or a 5 hectometer race.

For long distances . . . 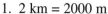 . . . as from city to city

or

cross country . . .

. . . the *kilometer* is used.

The kilometer is a measure of length equal to 1,000 meters. Its symbol is km. Kilometer is pronounced **kil** o meter, not ki **lom** eter. In fact, *the rule for pronouncing all metric units is to place the primary accent on the first syllable.* It would take about 15 minutes to walk a kilometer. The length of nine football fields placed end to end would be about one kilometer. Kilometers can be measured using the distance measuring gauge, called an odometer, in a car. Maps drawn to metric scale can help you determine long distances.

For practice in converting one metric unit to another, complete the following and check your answers below.

1. 2 km = _____ m   5. 2 dam = _____ m

2. 3 km = _____ m   6. 500 m = _____ hm

3. 4000 m = _____ km   7. 1 km = _____ dam

4. 21,000 m = _____ km   8. 1 km = _____ hm

Compare your answers with those in the Self-Check.

Self-Check _____ Activity 4.4 ✔

1. 2 km = 2000 m   5. 2 dam = 20 m
2. 3 km = 3000 m   6. 500 m = 5 hm
3. 4000 m = 4 km   7. 1 km = 100 dam
4. 21,000 m = 21 km   8. 1 km = 10 hm

Activity 4.5 Measuring Mass Metrically

First a word about mass and weight . . .

There is a very definite difference between mass and weight. Weight refers to how heavy an object is, while mass refers to how much stuff, or matter, is in an object. Weight is the result of the pull of gravity. If the force of gravity changes, then the weight of an object changes. On the moon, where gravity is less, you will weigh less than on Earth, but your mass remains the same. So if you want to lose weight, go to the moon, but if you want to lose mass, you will need to consider diet and exercise.

When we talk about our bone mass, or muscle mass, we do not care about how much our bones or muscles weigh, but rather we care about how much bone or muscle material is actually present.

The problem lies in that people often use the word weight when they really mean mass. Mass and weight in everyday life are being treated as if they were the same. Consequently, the kilogram is used to measure both mass and weight. For a more scientific approach to measuring weight due to gravity, see the section on *Measuring Forces.*

The metric base unit of mass is the kilogram. How much is a kilogram?

 . . . The mass of the water in a cube this size is 1 gram.

 . . . Hold a nickel in your hand and try to get a *feel* for its mass. Its mass is about 5 grams.

Masses larger or smaller than the gram can be described using the prefixes with the gram unit. The *kilogram* is the unit used in measuring most large masses. One kilogram is equal to 1000 grams and is the mass of one liter of water. An average man would have a mass of about 80 kg. Very large masses are measured in *metric tons.* One metric ton is the mass of 1000 kilograms.

Masses smaller than the gram are measured in *milligrams.* To get an idea of how small one milligram is, pick up a postage stamp and think about it having a mass of about 20 milligrams.

One way to find the mass of an object is to use an equal-arm balance, like the one pictured below. Place an object of unknown mass on one side and then balance it with objects of known mass on the other side.

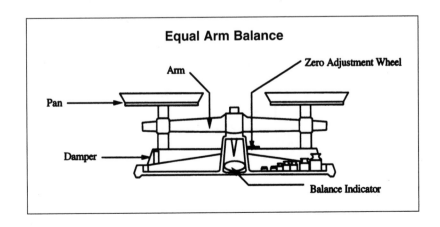

Before massing an object, always be sure the balance you are using is *zeroed*. This means that the two empty pans are in balance. If the balance is zeroed, the balance indicator will point to the exact center of its scale. If the indicator is off center, turn the *zero adjustment wheel* slightly until the indicator does point to the center of its scale. This zero adjustment wheel simply corrects for an out of balance condition. If the balance is not zeroed prior to massing, your results will not be accurate.

On the left side of the base is a small projection on some balances that resembles a paper clip. This is called a *damper*. By pushing and holding the damper the pans will stop moving.

Stored at the base of the balance are the *standard gram masses*. These masses are the objects of known mass with which objects of unknown mass are compared. Simply place the object of unknown mass in one pan, then add masses to the other pan until the pans balance and the indicator points to the center. The mass of the unknown is the total of the masses it takes to balance the pans.

Use the following procedure to practice massing objects with an equal-arm balance.

1. *Zero* the equal-arm balance; use the damper to stop the pans from moving if necessary.
2. If you are finding the mass of chemicals, *cover* the pan with a small piece of paper so that the balance is kept clean.
3. *Place the object or substance* you wish to mass in the center of one of the pans. The general rule or convention is to place the object in the left pan and the known masses in the right pan.
4. *Add masses* to the center of the other pan until the balance arms are horizontal.
5. *Total the masses* it takes to balance the unknown mass.
6. *Record* the observed mass.

→ GO TO the supply area and obtain an equal-arm (double pan) balance, 20 centicubes and a set of masses. Find the mass of 5 centicubes and record your finding in the *Your Masses* column of the chart below. Do the same for 20 centicubes. We have massed the same objects and recorded *Our Masses* in the chart so that you may have a check for your measurements. Slight differences may occur; you will need to exercise some judgment in comparing your masses with ours.

Object	Our Masses	Your Masses
5 centicubes	5 grams	4.5
20 centicubes	20 grams	19.7

Now that you know how to use an equal-arm balance, **go** to the supply area where you will find a container of soda, bag of potato chips, and container of cookies or other items provided by your instructor. Do not open any of the products. Look at the labels and record the amount (mass) of sugar and the amount (mass) of fat for each product in the table below. Be sure to use the appropriate units of mass. If a container holds more than just one serving of the product, calculate and record how much fat or sugar would be in the whole container. Do you usually pay attention to serving sizes when you buy a product?

Product Name	Mass of Sugar in Entire Container	Mass of Total Fat in Entire Container
Soda:		
Potato chips:		
Cookies:		

What do those amounts of sugar and fat look like? Return to the supply table where you will find a container of sugar and a container of solid shortening. For one or more of these products, measure an amount of sugar equivalent to the mass on the product label. Then, using the solid shortening to represent fat, measure an amount of shortening equal to the fat content found in the product. Share your findings with your classmates and think about what you found the next time you eat or drink one of these products.

 **Activity 4.6   Measuring Volumes Metrically**

In the metric system the *liter* is the base unit used to measure the volume of liquids. The symbol L is used for liter. A liter is the amount of substance that a cubic decimeter box can hold. 1 dm × 1 dm × 1 dm = dm³; a liter = dm³.

➡️**GO TO** the Metric supply area and pick up a liter container. Fill the container with water and try to get an idea of how much 1 liter really is.

➡️**GO GET** at least four other containers of different sizes and shapes, and estimate whether each would hold less than one liter, more than one liter, half a liter, two liters, and so on. Then check your estimates by pouring the water from the liter-measure into each of the containers. Refill the container as needed.

Container	Estimate	Measure
1 meds	less 30ml	40ml
2 Baby food jar	less 250ml	132ml
3 plastic cup	less 260ml	235ml
4 Jar	less 870ml	880ml

Again, prefixes are used to show very large or very small quantities. Because most of the substances that you will be measuring are relatively small, you will need to learn about *milliliters*.

The cube pictured is a centicube. A centicube measures 1 centimeter by 1 centimeter by 1 centimeter, so its volume is one cubic centimeter. If we could fill this centicube with liquid, it would hold 1 *milliliter* of liquid. The symbol mL is used to symbolize milliliter.

Any container that is graduated in milliliters can be used to measure small amounts of liquid.

→**GO TO** the supply area and obtain a grad-uated cylinder and pour some water into it. If the graduated cylinder is glass, you should notice that the upper surface of the water is curved or crescent-shaped. This curved surface is called the *meniscus.* When you measure the volume of a liquid, you should line up your eyes, as shown in the diagram, with the bottom of the meniscus. What volume is shown in the diagram to the right? If you are using a plastic graduated cylinder, it will not have a meniscus.

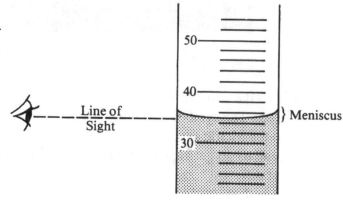

You should have read the volume as 35 mL (halfway between the 34 and 36 mL marks.)

As you have probably already learned, it is important to determine how many milliliters are represented by each mark on the graduated cylinder. It may differ from container to container. How much is contained in each of the following graduated cylinders?

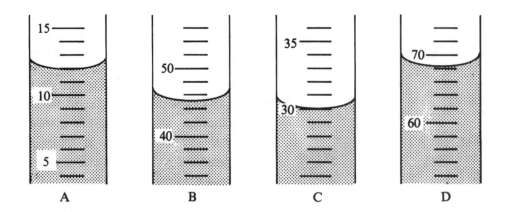

The answers are given below.

Self-Check ─────────────────────────────────── Activity 4.6 ✔

      a. 12 mL
      b. 45 mL
      c. 30 mL
      d. 68 mL

Although we use liters and milliliters to measure the volume of liquids, we use different units for meas-uring the volume of solids or of empty space such as rooms. The cubic meter is the unit used to measure large volumes of solids or space. The typical kitchen stove is approximately 1 meter by 1 meter by 1 meter, or m^3. When measuring smaller solid objects and substances, we use the cubic centimeter. Cubic centimeter is sym-bolized as cm^3. The old abbreviation of 'cc' for cubic centimeter is now considered incorrect and should be avoided.

Suppose you wanted to find the volume of a small rock. You would need to determine two things:

1. The procedure you would use to measure the rock.
2. The units you would use for your answer.

Because the rock you want to measure is solid and relatively small, the appropriate unit of measurement is the cubic centimeter (cm^3). If the rock is irregular in shape, it would be impossible to measure its volume accurately with a ruler or measuring tape. But if you gently placed the rock in a graduated cylinder of water, you would notice that the water would rise in the cylinder. The volume of the rock would be equal to the amount of water displaced. If the water rose 4 mL, what would be the volume of the rock? You will have the answer as soon as you know how a milliliter and a cubic centimeter are related. Do this simple activity to find out:

Fill a graduated cylinder with some water. Note how much water is in the cylinder. Place 1 centicube in the graduated cylinder and note the rise in the water level. You should see a 1 milliliter rise in water level. If you recall that a centicube is a cubic centimeter, try writing a statement that relates millimeters (mL) and cubic centimeters (cm^3).

Recall that a liter is equal to a cubic decimeter (dm^3). A dm^3 is a box that is 1 dm × 1 dm × 1 dm. One decimeter is equal to 10 centimeters, so a liter is also equal to a box that is 10 cm × 10 cm × 10 cm, or $1,000 \ cm^3$.

$$\text{Liter} = dm^3 = 1,000 \ cm^3$$

One liter is equal to 1,000 milliliters; recall that milli means 1/1000, so 1,000 x 1/1,000 = 1

$$\text{Liter} = dm^3 = 1,000 \ cm^3$$
$$\text{Liter} = 1,000 \ ml$$
$$\text{Therefore, } 1,000 \ cm^3 = 1,000 \ ml, \text{ so } \mathbf{1 \ cm^3 = 1 \ ml}$$

What if the rock whose volume you want to measure is too big to fit in a graduated cylinder? Then you can use what is called an *overflow container* to determine how much water your rock displaces. An overflow container is simply a container with a spout. To use the overflow container, fill it with water until water overflows through the spout. When the water stops overflowing, place the object whose volume you want to measure into the container. The object will displace an amount of water equal to its own volume and that amount of water will flow out the spout where it can be collected and measured. (If the object floats, you will need to push it under water.)

If you do not already have an overflow container, here is an easy way to make one:

Optional Activity 4.7 Making an Overflow Container

You will need:

1 clean 2-liter bottle with the label removed

1 standard 6mm diameter plastic, bendable straw

1 standard round, 6 mm diameter, paper hole punch

1 pair of scissors

1. Prepare the bottle: Remove the top of the bottle as shown, cutting just below the shoulder of the bottle. Use a hole punch to make a single hole on one side of the container, about 2 cm below the top edge.
2. Prepare the straw by cutting right next to the accordion elbow of the straw so that it is the long, straight shank that you cut off. It is the short piece of straw with the elbow at one end that you will use.
3. Assemble the parts. Hold the straw by its elbow and push the straight part of the straw through the hole in the bottle *from the inside out.* The elbow secures the straw in the hole and the straw sticks out to form the spout.

For practice, try the following.

➡️ **GO TO**   the supply area and obtain a solid rock, a porous rock, a graduated cylinder and an overflow container. Find:

a. volume of a solid rock
b. volume of a porous rock
c. volume of your hand.*

Self-Check _____ Activity 4.7   ✔

> Compare your answers with someone else's. Check to be sure you recorded the volumes of the liquids in liters or milliliters and the volume of solid objects in cubic centimeters.

 Activity 4.8   **Measuring Temperature Metrically**

For everyday purposes, temperature in the metric system is measured with a Celsius thermometer. Examine the Celsius thermometer pictured on the following page. Observe the three standard temperatures marked on the scale: the temperature at which water boils (100 °C), the temperature at which water freezes (0 °C), and normal human body temperature (37 °C).

*Here's a challenge question for you: How might a healthcare worker measure and monitor the amount of swelling in a person's injured hand?

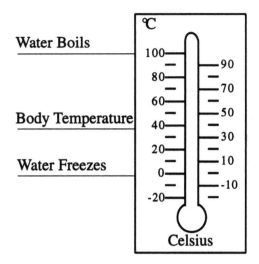

The symbol ° means *degree*. The temperature 20 °C, for example, should be read as twenty degrees Celsius.

➡️**GO TO** the supply area and obtain a Celsius thermometer. Examine the scale on the thermometer. Is the thermometer calibrated in one degree intervals? Two degree intervals? Five degree intervals? Measure the room temperature.

1. How many degrees are represented by each interval of the scale?
2. What is the room temperature?

Check your answers with someone else's.

Obtain 1 small beaker or similar container and some ice from the supply area. Keep the Celsius thermometer. Fill the container half full of cold water. Measure the temperature of the water. Be sure to give the thermometer a few seconds to adjust before reading it. Add an ice cube to the water. Measure the temperature every two minutes while stirring the water gently. Complete the following table:

Time	Temperature
Temperature of water before adding ice	
Temperature of water after adding the ice: After 2 minutes	
4 minutes	
6 minutes	
8 minutes	

Compare your answers with the ones that follow.

Self-Check _____ Activity 4.8

Your table should be like this:

Time	Temperature
Temperature of water before adding ice	18°C
Temperature of water after adding the ice: After 2 minutes	13°C
4 minutes	12°C
6 minutes	10°C
8 minutes	10°C

Your answers will be different from ours. The initial temperature of the water and the room temperature are two factors that will influence the results. Can you think of other factors that might explain why our results differ?

Activity 4.9   Measuring Forces Metrically

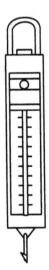

Whenever you measure a push or pull, you are measuring force. The unit of force in the metric system is the *newton* (N). Newtons are a measure of how much force is being exerted on an object.

Instruments like the spring scale shown on the left and the personal bathroom scale are used to measure force. The spring scale may be hung, held, or laid on a flat surface. The greater the force being measured, the farther the spring is stretched. Objects that cannot be hung on the scale's hook may first be placed in a mesh bag, such as an onion bag, that can be hung on the hook.

In addition to measuring the pull of gravity upon objects, spring scales can be used to measure (in newtons) numerous other forces such as the force it takes to pull an object up an inclined plane. A simple instrument for measuring small forces could be made using a rubber band. Hang objects on the rubber band and calibrate the distance each stretches the band.

Weight is a measure of the force that results from the pull of gravity on nearby objects. The Earth's gravitational pull on objects is stronger than the moon's gravitational pull on objects near its surface. People would weigh about one sixth of their Earth weight on the moon.

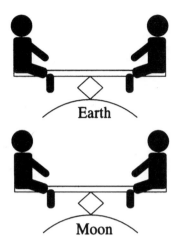

Earth

Moon

Remember that mass is the measure of the amount of matter in an object. Although weight changes due to changes in the pull of gravity, the mass of an object would remain the same wherever it is placed in the universe.

Two children balancing one another on a teeter-totter (or see-saw) on Earth would still balance one another on the moon. Each child's *mass* would be the same on the moon as on the Earth, but each child would *weigh* about one-sixth as much on the moon as on Earth.

If we keep our thoughts Earth bound, we can state a definite relationship between mass and weight. At sea level, on the Earth's surface, a kilogram mass weighs about 10 newtons.

Practice measuring forces. At the same time you will be visiting some really big ideas in science:

➡**GO TO** the supply area and obtain a small rock, a spring scale, and mesh bag. Find the rock's weight by (1) placing it in the bag; (2) hanging the bag from the spring scale's hook, and (3) reading the scale. Weight is the measure of the force of gravity acting on the mass of the rock.

Record the weight of the rock: _____.

Gravitational force is one of the big ideas in science that we want students to experience.

Repeat the procedure you just used, but this time lower the bag with the rock in it (but not the scale) into a container of water. Do not get the scale wet and do not let the rock touch the bottom of the container. Read the scale. Does the rock *appear* to weigh less in water?

Record the *apparent* weight of the rock as it is buoyed up by water: _____

Using the measurements you just recorded (the rock's actual weight and its apparent weight in water), determine the buoyant (upward) force of water acting on the rock. _____

Buoyancy force is another big idea in science.

With the spring scale still attached, place the bag with the rock on a table or other flat surface. As you drag the scale slowly pulling the rock across the surface, watch the scale carefully. Record how much force is needed to *just get the rock moving* from its stationary position: _____

Force needed to overcome inertia is also a big idea in science.

Once you get the rock moving across the surface, how much force is needed to *keep it moving* at a steady rate? Record the force needed to keep the rock moving: _____

How does the force required to keep the rock moving compare with the force necessary to initially set it in motion? _____

Self-Check _____ Activity 4.9 ✔

Make sure all the forces you measured were in newtons (N).

✔

Continue on and then take the Self-Assessment that follows.

 **Self-Assessment**   Measuring Metrically

I. In the space provided, write the most appropriate metric unit in which each of the following should be measured:

1. The length of a basketball court should be measured in _____.

2. The temperature of water should be measured in _____.

3. The volume of an orange should be measured in _____.

4. The diameter of a pencil should be measured in _____.

5. The mass of a refrigerator should be measured in _____.

6. The width of a VCR should be measured in _____.

7. The amount of water used in a baking recipe should be measured in _____.

8. The amount of gas an automobile gas tank can hold should be measured in _____.

9. The mass of a package of chewing gum should be measured in _____.

10. The mass of a postage stamp should be measured in _____.

11. The weight (due to gravity) of a shot an Olympic shot putter throws should be measured in _____.

12. The mass of a shot an Olympic shot putter throws should be measured in _____.

II. Use what you know about metric prefixes and state equivalent metric measures for each of the following:

1. 1 meter = _____ decimeters

2. 1 dm = _____ cm = _____ mm

3. 1 mm = _____ m = _____ dm = _____ cm

4. 2 kilometers = _____ meters

5. 4000 m = _____ km

6. 500 meters = _____ hectometers

7. 1 km = _____ dam

8. 1 km = _____ hm

9. 1 liter = _____ milliliters

III. ➡**GO TO** the supply area and obtain the *Measurement Test* in the sealable plastic sandwich bag and do the following:

1. Measure the height of the sandwich bag to the nearest half (0.5) centimeter.

    bag height: _____

2. Measure the width of the sandwich bag. Express your measurement in millimeters, centimeters, decimeters, and meters.

    bag width: _____ mm _____ cm _____ dm _____ and m _____

3. Fill the small plastic cup with water and measure the temperature of the water.

    water temperature: _____

4. Measure the volume of water the small plastic cup can hold.

    water volume: _____

5. Measure the volume of gravel the cup can hold.

    gravel volume: _____

The answers to the Self-Assessment follow.

Self-Assessment Answers

I. 1. length of basketball court—*meters (m)*
    2. temperature of water—*degrees Celsius (°C)*
    3. volume of an orange—*cubic centimeters (cm³)*
    4. diameter of a pencil—*millimeters (mm)*
    5. mass of a refrigerator—*kilograms (kg)*
    6. width of a VCR—*centimeters (cm)*
    7. amount of water used in a recipe—*milliliters (mL)*
    8. amount of gas a tank can hold—*liters (L)*
    9. mass of a package of gum—*grams (g)*
    10. mass of a postage stamp—*milligrams (mg)*
    11. weight (due to gravity) of the shot an Olympic shot putter throws—*newtons (N)*
    12. mass of the shot an Olympic shot putter throws—*kilograms (kg)*

II. 1. 1 meter = 10 decimeters
     2. 1 dm = 10 cm = 100 mm
     3. 1 mm = 1/1000 m = 1/100 dm = 1/10 cm
     4. 2 kilometers = 2000 meters
     5. 4000 m = 4 km
     6. 500 meters = 5 hectometers
     7. 1 km = 100 dam
     8. 1 km = 10 hm
     9. 1 liter = 1000 milliliters

III. 1–5. Check your answers with those given *on the card* in the *Measurement Test* bag.

IDEAS FOR YOUR CLASSROOM

Measuring for the sake of measuring is dull and pointless, so incorporate measuring with other class activities. Here are some suggestions upon which you can expand with your own ideas.

The activities you choose should depend a great deal on your students' interests.

1. Construct other measuring instruments from the ones already available in the classroom. What would you need to measure the distance between the office and the cafeteria? The distance around your waist? The distance from the second story window to the ground?
2. Measure growing plants. Keep a record and make comparisons between plants grown in different conditions.
3. Construct a map of the classroom representing actual distances.
4. Measure shadows at different times of day.
5. Keep a record of the amount of food and water a classroom pet requires each day.
6. Here is a (metric) recipe for growing crystals. The results are fascinating!
    5 mL household ammonia
    15 mL water
    15 mL table salt
    15 mL bluing
    Mix together and pour over rocks, sand, sponges, wood, or bits of brick and cement. Spread out on a tray. Let it stand and watch the crystals grow.
7. Make bread, cookies, or pudding. Measure the ingredients, temperature, and time.

Before you measure, what is your estimate?

What is this object's length, mass, volume, area, and so on?

Quantify that observation.

Use numbers to describe what you observe about this object or event.

Measure and record what you observe about this object or event.

How long (or heavy) do you think this object is?

High Stakes Testing

A sample multiple-choice item from State Standardized Exams.

You want to measure the volume of a rock. The best way to do this is with
- A. *a ruler, water and a balance.*
- B. *a beaker and water.*
- C. *a beaker, a hot plate and water.*
- D. *a beaker, a graduated cylinder and water.*

Oregon: Benchmark 3: Sample Text 2001–02.

Technology Spotlight

Using Search Engines Effectively

The Internet is a wonderful tool for both teachers and students. On the Web you can easily find thousands of lesson plans, detailed science content information, and numerous interactive student resources. A search, however, can often lead to an overwhelming number of sites. Having two skills will help you receive a manageable amount of information that meets your needs: 1. selecting appropriate search engines, and 2. using the search features of those engines to quickly locate specific information and resources.

Some popular search engines used by educators are Alta Vista, Northern Light, and Google. A useful site to help you locate and evaluate different search engines is Beaucoup (http://www.beaucoup.com). This site lists more than 2,000 search engines, indices, and directories, and organizes them into categories. In addition it provides reviews and tips for using many of the most popular search engines.

Each search engine has its own set of rules for focusing a search that are accessible through the help or advance search buttons. Knowing how to use these advanced search options is what enables the Internet to become an indispensable tool for teachers. For most search engines you will find some common searching functions, such as domains, and Boolean and proximity operators. In addition many search engines can be searched by phrase or by title. Image searches are also possible, but images on the Web follow the doctrine of applied public access—images should not be used without explicitly stated permission. The option to conduct searches that are filtered for objectionable materials is also available on most search engines, which is a very helpful feature for searches conducted in school settings.

http://oncampus.richmond.edu/academics/as/education/projects/webunits/measurement/home.htm
http:www.muohio.edu/dragonfly/small/
www.sasked.gov.sk.ca/docs/elemsci/g2fslc5.html
www.aaamath.com/mea.html
www.monroe2boces.org/shared/esp/measure.htm
tqjunior.thinkquest.org/3804/

WEBSITES FOR ACTIVITIES

A Model for Assessing Student Learning

Assessment Type: Performance Task

Preparation

You will need to set up labeled stations around the room with the materials listed below for each station. Manage the flow of students by having some students start at different stations. If the room size allows, have more than one set of stations.

Directions to the Students

_____ 1. Go to each station set up in the room (as directed by your teacher). For each station follow the directions given on this page.

_____ 2. Answer the questions for each station in the space provided on this page.

_____ 3. When recording your measurements, record both the number and the unit of measurement (for example, 10 centimeters).

Station A

Check your materials: metric ruler and meter stick.

_____ 1. Measure the width of this paper in centimeters.

_____ 2. Measure the length of this line in millimeters.

_____

_____ 3. Measure the length of this tabletop in meters.

Station B

Check your materials: graduated cylinder, plastic cup with a black line, container of water.

_____ 4. Pour water into the cup until it reaches the black line. How much water is in the cup?

When you are finished, pour the water back into the container.

Station C

Check your materials: dual scale thermometer.

_____ 5. What is the room temperature in degrees Celsius?

Station D

Check your materials: two-pan balance, set of masses, object to be massed

_____ 6. What is the mass of the object?

When you are finished, remove the masses and the object from the balance.

Chapter 5

Inferring

National and State Standards Connections

- Develop explanations using observations (evidence) and what they already know about the world. (NSES K–4)
- Discuss events related to students' experiences as likely or unlikely. (NCTM, Pre-K–2)
- Distinctions are made among observations, conclusions/inferences, and predictions. (Virginia: Standards of Learning, Grade 4)

MATERIALS NEEDED

- ✔ clean, empty soda or juice can
- ✔ access to the Internet
- ✔ small balloon
- ✔ piece of soft cloth
- ✔ a tree cross section (optional)

- ✔ bar magnet
- ✔ 20 cm string
- ✔ ruler
- ✔ mystery box

 **Classroom Scenario**

Mr. C, a fourth grade teacher, came to class one day carrying an old, worn athletic shoe. He held the shoe up so all the students could see it.

Mr. C: This morning I found this shoe in the road next to the school. How do you think it got there?

Michael: It's ugly. Somebody threw it there.

Mr. C: That's one possibility. What are some other ways it might have gotten there?

Devon: Somebody ran across the road and their shoe came off.

Dillon: It fell out of somebody's backpack.

Alicia: It fell out of somebody's car.

Mr. C:	How else might it have gotten there?
Cally:	Maybe somebody got hit by a car crossing the road and their shoe fell off!
Mr. C:	Do we have more ideas about how it might have gotten there?
Maria:	Maybe it fell off a garbage truck. Does it stink?
Sally:	Maybe it was a dog's toy and the dog dropped it in the road.
Mr. C:	Those are all great ideas. I like the way Maria asked a question that we can answer by investigating further. Asking good questions can help us learn more about what actually happened. What are some other questions we could ask and investigate?
Donny:	Yeah, does it stink?
Cally:	Is there blood on the shoe?
Sam:	Does it have teeth marks?
Devon:	Is somebody missing a shoe?

(The class then tried to answer their own questions by making closer observations.)

Mr. C:	We've been able to answer some of your questions by making closer observations. Let's see what we can conclude from what you observed. Cally, do you still think there was an accident?
Cally:	Maybe not, we didn't see blood stains.
Mr. C:	Donny, do you think it was in the garbage?
Donny:	It could have been if it was in clean garbage.
Sally:	And we didn't find any teeth marks, so maybe there wasn't any dog.
Donny:	OK, so what's the right answer? How DID the shoe get there?
Mr. C:	We don't know for sure.
Georgia:	You mean we did all this for nothin' and we'll never know?
Mr. C:	We have no way of knowing for sure because we cannot go back and observe what actually happened. We can only infer or conclude what we think happened based on good observations and other things we know. Now we know more than we did before because we asked good questions and tried to answer them. That is what scientists do when they have questions and try to find answers.
Donny:	Oh, maaan!

───────────────── Classroom Scenario ─────────────────

In this scenario children were making observations, gathering information, and trying to interpret and explain something they are curious about. These are the same things scientists do when they investigate to find answers to their questions. Scientists are curious about things such as: *How do we catch a common cold?, What causes house fires?, Why do large numbers of fish sometimes die in a lake?, Did asteroids at one time interfere with life on Earth?, How can buildings be constructed to withstand earthquakes?, What causes cancers and birth defects, and How can plants can be grown to resist disease?* It would be nice if answers to these questions, and others like them, were simple. The human mind seems to like single "right" answers, but things are seldom that simple, in part because few events have just one cause. Scientists know, for example, that the common cold is caused by a virus. Not every person exposed to the virus, however, actually gets a cold.

Finding that there may be alternative answers, or no immediate answers at all, can be uncomfortable, confusing, and frustrating for scientists and children alike. We can help children to better understand their world by arousing their curiosity and encouraging them to ask questions, make good observations, look for patterns, and organize information.

We can think of children as theory builders. As they experience their world and are curious about things in it, they develop their own explanations (theories) about how things work. Or if they are not interested, they may make no attempt to explain things at all. The theories that children build early in life are often the same theories they carry into adulthood. These 'naïve theories' may or may not be consistent with those that scientists believe about how things work. The belief that magnets have magical powers, for example, is not consistent with what scientists think. Scientists sometimes change their minds about what they believe as they continue to observe, gather new information, and try to make sense of new findings. Science, therefore, is not a static body of knowledge.

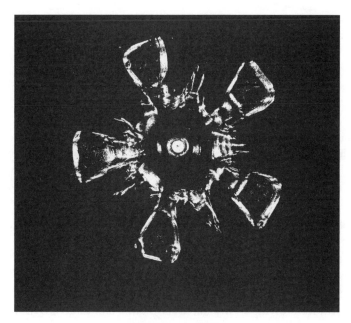

What is it?

Goals

In this chapter you will develop the skills necessary to make inferences based on observations.

Performance Objectives

After completing this set of activities you should be able to:

1. Given an object or event, construct a set of inferences from your observations about that object or event.
2. Given additional observations about the object or event, identify the inferences that should be accepted, modified, or rejected.

Constructing Inferences from Observations

While an observation is information perceived through one or more of the senses, an inference is an explanation or interpretation of an observation. In making an inference we use information already known from past experience and new information we directly observe through our senses.

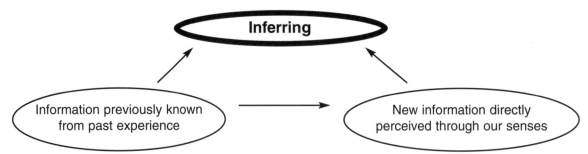

An inference is a statement that attempts to interpret or explain a set of observations. It follows that every inference must be based on one or more observations. An inference is not a guess because guesses are often based on little or no evidence. Use the following examples to help you distinguish observations from inferences. Each of these examples includes an observation that is followed by an inference that attempts to explain that observation.

- The brass knob on that office door is not bright and shiny. I infer that the office is not used often.
- There is a spot in my front yard where grass does not grow. Someone may have spilled a toxic substance there.
- I see that iodine turns purple when I put it on a potato chip. It can be inferred that the chip has starch in it.
- The pages in this book are yellow. I infer either that the book is old or that the paper was dyed yellow to give it an old appearance.
- Through the window I see the flag waving. It must be windy out.
- The fish are floating on top of the tank. Perhaps no one fed the fish.
- My drinking water smells like rotten eggs. Maybe it has become contaminated.
- The cabbages that were growing in my garden are gone and there are droppings on the ground. That is evidence that rabbits have been there.
- That star is brighter than the others. I infer it is closer to Earth than the others.

When making inferences, it is helpful to follow these steps:

1. Make as many observations about the object or event as possible.
2. Recall from your experiences as much relevant information about the object or event as you can and integrate that information with what you observe.
3. State each inference in such a way that clearly distinguishes it from other kinds of statements, such as observations or predictions:

   "From what I observe I infer that . . ."
   "From those observations it can be inferred that . . .
   "The evidence suggests that . . . may have happened."
   "What I observe may have been caused by . . ."
   "A possible explanation for what I see is that . . ."
   "From what I observe I conclude that . . ."

In Activity 5.1, *Constructing Inferences from Observations,* you will be constructing some inferences of your own. Remember that an observation is information perceived through one or more of the senses. An inference is an explanation or interpretation of an observation. The words conclusion and concluding are often used as synonyms for inference and inferring. This is particularly true in the various ways state science standards are written.

Activity 5.1 Constructing Inferences from Observations

→GO TO the supply area and obtain a clean, empty soda can. Examine it carefully. Do the following:

1. Make as many observations as you can about the can (remember to describe properties) and list them in the observation column below.
2. Try to think of things you already know that might help to explain or interpret what you observe about the can.
3. In the inference column list as many inferences as possible about the can, being sure to state each inference in such a way that indicates that it is an inference.
4. *Draw a line* connecting each inference to the observation on which it is based. (More than one inference may be inferred from one observation.)

Just to help you get started, here are a couple of possible observations and inferences someone else might make about his or her can:

Observations	Inferences
The can has several colors on it. ————	I infer it was painted.
The can does not have a seam along its side.	I infer that it was punched out of a flat sheet of material.
	Or it could have a very fine seam that I cannot see.

Soda Can

Observations	Inferences

Self-Check _____ Activity 5.1

The following are some observation and inference statements that someone else made about the soda can. Because your can and your own past experiences are different, your observations and inferences will differ. The Self-Check will help you to decide if you applied the skills of observing and inferring properly. You may have phrased your inferences slightly differently.

Soda Can

Observations	Inferences
The bottom of the can is a silver color.	I infer it is made of aluminum.
The can has "12 FL. OZ. (355 mL)" on it.	The can probably held that much liquid.
This can has a raised tab on the top.	I infer someone used it to open the can.
There is a brown substance on the rim.	Perhaps the substance came from inside the can.
When I squeeze the can, it makes a "clinking" sound.	I infer the material is thin.
There are a lot of little dents on the sides.	I infer the can came out of a dispensing machine.
The can has the word "Diet" on it.	Maybe the person who drank the soda wanted to lose weight.

Activity 5.2   More Practice Making Inferences about Observations

➜**GO TO** this Web site: http://www.phoenix5.org/gallery/earthlights1500.html

Or you may search for this site by using "NASA and Earthlights" as your search phrase, then click on "Main Menu for EarthLights, NASA photos of lights of Earth at night."

What a spectacular image of light on Earth! Examine the image carefully and do the following.

1. Make as many observations as you can about the image and list them in the observation column on the following page.
2. Try to think of things you already know that might help to explain or interpret what you observe about the image.
3. In the inference column list as many inferences as you can about the image, being sure to state each inference in such a way that indicates it is an inference.
4. Connect each inference with the observation on which it is based.

Earth at Night

Observations	Inferences

Self-Check _____ Activity 5.2 ✔

Compare your observations and inferences with someone else's.

Activity 5.3 **Formulating Inferences from Observations about Events**

This next activity is designed to help you learn to formulate inferences about events. Every inference must be based on an observation, so you will first be making careful observations and then interpreting or explaining those observations. These interpretations or explanations of observations are inferences.

➡ **GO TO** the supply area and obtain a small round balloon, a piece of soft cloth, and a clean, empty soda can. Make sure the can you use is not dented. Follow these steps to make an interesting event happen. Carefully observe everything that is taking place as it happens.

1. Fill the balloon with enough air to make its surface fairly tight.
2. Clear an area on a flat desktop or table where the can will be able to roll back and forth a short distance.
3. Place the can on its side, then avoid touching it again with your hands.

4. Rub the balloon gently on the cloth a few times and bring the balloon **near** (about 2 or 3 cm) but **not touching** the can's surface. You should see the can begin to roll. (If bringing the balloon close to the can does not cause the can to roll, consult your instructor about factors such as room humidity that may be affecting the results.)

5. Try controlling the can's back and forth movements just by moving the balloon from one side of the can to the other.

6. Now that you have witnessed the event, do the following:

   a. Record your observations in the observation column of the chart below.

   b. In the inference column list as many inferences as you can about the event and state each inference in such a way that indicates it is an inference. Try to think of things you already know that might help to explain or interpret what you observed about the event.

   c. Draw a line connecting each inference with the observation on which it is based.

Rolling Can

Observations	Inferences

Self-Check _____ Activity 5.3 ✔

Compare your observations and inferences with someone else's.

Activity 5.4 Making Inferences from a Slice of Life

→ **GO TO** the supply area and obtain a tree slice (a cross section of a tree) if one is available. If not, use the illustration accompanying this activity.

1. Carefully examine the tree cross section or picture. Record as many observations as you can in the chart below.
2. Recall what you already know about trees. Trees need light, air, water, minerals, and space in which to grow. You may already know that each year a tree grows it produces new wood tissue which, in cross section, appears as concentric circles, called growth rings. Spring growth tissue and summer growth tissue differ in texture and color. A darker ring together with a lighter ring represents one year's growth. When growing conditions are favorable, rings are wide. When growing conditions are not favorable, rings are narrow. If you are having difficulty seeing the growth rings on your own sample, try wetting the surface or using a magnifying glass. Think of some things you already know that might encourage tree growth or inhibit it.
3. Using what you observed and what you already know, make some inferences about your tree's age, health history, and surrounding environmental conditions. Record your inferences in the inference column in the chart.

Tree Cross Section

Observations	Inferences

Self-Check _____ Activity 5.4

The inferences you made will depend on the tree cross-section that you used as well as your past knowledge about trees. In making your inferences, you may have considered the following.

If the tree's growth rings are narrow and close together, growth may have been slowed by such things as drought, low light conditions due to cloudy weather or a nearby structure such as a building, poor soil conditions, crowding from other tress, low temperatures, harmful insect infestation, or disease.

If the growth rings are wide and far apart, the tree's growth may have been enhanced by plenty of rainfall, by sufficient amount of light, good soil, and warm temperatures, and by the absence of disease and harmful insects.

The other inferences you may have made depend upon what observations you were trying to explain. Were the tree rings the same distance apart all around the tree? Were there holes in the bark? Was the inner part darker than the outer part? Are there additional lines radiating from the center out to the bark?

Learning Is an Inference

Recall that an inference is an explanation or interpretation of what you observe. When you are inferring, you try to give meaning to what you observe. Inferences, and hence new learning, occur when you make sense of your observations.

Because inferences are based not just on what is observed but also on what the observer already knows, it follows that new experiences may be interpreted differently by different people. Each person constructs his or her own learning depending upon past experiences and knowledge already gained. Not everyone walks away from the same experience having constructed the same new knowledge. Therein lies a problem, particularly for teachers. For children to learn something new they must relate the new concept to concepts they have already formed. If an individual child is unable to relate the new concept to any pattern of knowledge he or she already has, learning does not take place. The teacher's role is to facilitate learning by linking new concepts with what individual children already know.

You, as a teacher, must be an expert observer and questioner. In order to link new knowledge to old knowledge, you will need to find out what that previous knowledge is. You can not assume it exists for all students. By creating the right situations, you can observe what individual children know and do not know. When prerequisite knowledge does not exist, your role as a teacher is to design activities in which children can use their senses to experience as much about a concept as they can. That experience base will in turn allow them to *construct* more knowledge.

Thinking Is All about Asking Questions

You can promote thinking by asking:

What do you observe?

What do you already know?

How can you explain what you observe?

What do you think it means?

You can promote *deeper* thinking by asking:

What do you observe? What *else* do you observe?

What do you already know? What *else* do you already know?

How can you explain what you observe? How *else* can you explain what you observe? What *else* might it mean?

What new information do you need? How can you get it?

You teach children how to think for themselves by encouraging them to *ask themselves* similar questions:

What do I observe? What *else* do I observe?

What do I already know about this? What *else* do I already know about this?

How can I explain it? How *else* can it be explained? What *else* might it mean?

What new information do I need? How can I get it?

Classroom Scenario

Mrs. Q took her class on a walk on the school grounds. The lesson's purpose was for the children to look for living things and to find evidence that other living things had been present. Finding such things as trees, grass, birds and ants was easy as those things could be observed directly. The presence of other living things could only be inferred from objects the children found, like the pine cones the children thought had been chewed by a squirrel, the dry skin casing they thought probably came from an insect, and so on. One group of children had this conversation with their teacher:

Quint: Look at what I found. These are funny looking leaves. They're full of holes.

Mrs. Q: Those are interesting. Where did you find them, Quint?

Quint: Over there on the ground. Look, you can see right through them.

Miguel: I found some too. They were kind of stuck to the ground.

Mrs. Q: What else do you observe about the leaves?

Miguel: They don't have any skin.

Mrs. Q: What do you mean, 'skin'?

Quint: You know, the green part. Other leaves have a green skin on them.

Mrs. Q: What else do you see?

Sara: Some have yucky black stuff on them.

Quint: And yucky white stuff too.

Mrs. Q: Have you seen that black and white 'yucky stuff' anywhere before?

Sara:	Maybe on moldy old bread.
Mrs. Q:	What are some possible reasons why the leaves look like that?
Sara:	Maybe the moldy stuff made the leaves sick.
Mrs. Q:	That is good thinking, Sara. Maybe that is mold on the leaves and that helps to explain why it looks the way it does and Quint, you said you found the leaves stuck in the soil. Think of some things you know live in the soil. What else might have happened to the leaves to make them look like that?
Dale:	Bugs and other stuff that live in the dirt might have eaten them and left the bones.
Mrs. Q:	What do you mean 'bones'?
Dale:	You know, when you eat stuff and leave the bones.
Miguel:	He means moldy stuff and bugs and other things in the dirt ate the soft part and left the hard parts.
Mrs. Q:	That is a good explanation for now. Let's look for other evidence that things might live in the soil even though we cannot see them.

––––––––––––––––––––––––––––––––– Classroom Scenerio –––––––––––––––––––––––––––––––––

Confidence in Inferences

Patterns develop from similar or related past experience. Observations that can be linked to patterns of known information form a basis for making inferences. If you observe something new that matches a pattern of information you already have, then you have considerable confidence that your inference really does explain what you observed. On the other hand, you may have little confidence in your inferences when observations do not match a preexisting pattern.

Just as in previous activities, in the following activity you will make both observations and inferences. This activity, however, differs in that you will also be asked to indicate how confident you are in what you infer. When making your inferences, be sure to use only words and ideas you understand. Your confidence in some inferences about this activity is likely to be high when your observations fit into a pattern of experience you already possess. When observations do not match a previously learned pattern very well, your confidence in accurately explaining what you observe is apt to be much less.

Take another look at the picture on page 83 of this chapter. Are you confident you know what it is? (Hint: It is a photocopy of a common object.)

Accepting, Rejecting, and Modifying Inferences

Often, after having drawn inferences or conclusions from a set of observations, new information becomes available that may cause you to rethink your original inferences. Sometimes additional observations reinforce your inferences. At other times, however, additional information may cause you to modify or even reject inferences that were once thought to be useful. New observations lead to adjusting patterns of experience to accommodate the new information. The science processes of observation and inference might look something like this.

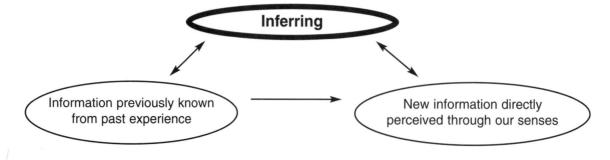

In science, inferences about how things work are continuously constructed, modified, and even rejected on the basis of new observations.

In the following activity you will make some initial observations about an event and draw some inferences from those observations. You will then have the opportunity to make additional observations. Your new observations will cause you to either accept, modify, or reject your first set of inferences. When you have finished, you will be asked to write one more inference, called a *conclusion*, which is a summary of all the inferences you decided to accept.

Activity 5.5 Compass Direction and Magnetism

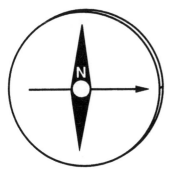

→**GO TO** the supply area and pick up a magnetic compass. Later you will need a bar magnet, a piece of string (about 20 cm long) and a ruler, but do not get these other items now.

Set the compass on the table in front of you and move it around a bit making sure there are no other compasses or magnets in the immediate area. Record what you observe about the behavior of the compass needle and where it points. Write down as many inferences as you can about what you observe.

Observations	Inferences

→**GO TO** the supply area and obtain a bar magnet, a piece of string (about 20 cm long), and a ruler. Use the ruler to find the magnet's exact middle and tie one end of the string tightly around the magnet at this midpoint. Attach the other end of the string to a stable object so the magnet hangs like a pendulum and can swing freely. A shelf, drawer handle or leg of a chair that has been turned onto its side will do just fine. Adjust the string slightly, if necessary, so the magnet is parallel with the floor.

Rotate the bar magnet slightly, then let it come to a rest by itself. Record what you see happening. Compare how the bar magnet has oriented itself with the orientation of the compass needle. Keep the magnet and compass at least 1 meter apart from one another; otherwise they will interfere with one another. Write as many inferences as you can to explain what you observed.

Observations	Inferences

 When the hanging bar magnet is at rest, bring the magnetic compass close to it. Move the compass under, over, and all around the suspended bar magnet. What happens? Record what you observe, and again make as many inferences as you can about what you observe.

Observations	Inferences

Drawing Conclusions

Activity 5.5 was related to magnetism and may have reinforced what you already knew about magnetism or may have caused you to adjust your thinking (patterning) about what magnetism is. After all, you can not see magnetism—you can only see what it does. To find out what adjustments, if any, you have made in your thinking about magnetism, do the following:

1. *Review* all the inferences you have made in this activity and accept, modify, or reject them by following these steps:
   - If you made inferences that you think are no longer good explanations for what you observed, reject them by drawing a line through them.
   - If you made inferences that you think are still OK but need some slight modifications to adequately explain what you observed, change them.
   - If you made inferences that you still feel are good explanations for what you observed, *circle* them to show that you accept them.
2. *Draw a conclusion:*
   - Write a statement below that *summarizes* what you were able to infer from all of your observations.
   - Make your inference statements simple and use only words that are meaningful to you. You are trying to explain phenomena in the context of what you already know.

From my activities about magnetism I conclude that:

_____

_____

_____

_____

_____

Of course, what you concluded will always be subject to revision and modification as your experience grows.

Self-Check _____ Activity 5.5 ✔

Here are some conclusions other students have drawn from doing the activities related to magnetism.

- *Something* is present in the environment that has the same effect on a compass that a magnet does.
- The Earth behaves as if it were a giant magnet.

 # Self-Assessment

Inferring

➔ GO TO the inference supply area and pick up a mystery box. Do not open the box or in any way tamper directly with the contents. Make at least five observations and five inferences about the contents of the box. Identify the specific observation on which you base each inference by drawing a line between them. You may tilt, shake, roll, or rattle the box but do not peek inside. List your observations and inferences in the chart below.

Observations	Inferences
1.	
2.	
3.	
4.	
5.	

Now open one end of the box and without looking inside put your hand in the box and gather some additional information about the contents. In the chart below, accept, reject, or modify each of your original inferences on the basis of this new information. Identify the observations on which you accept, reject, or modify each inference.

Observations	Inferences
1.	
2.	
3.	
4.	
5.	

Compare your observations and inferences with your partner's or check your answers with your instructor.

IDEAS FOR YOUR CLASSROOM

1. *Pictures* are excellent for use in developing skills of observation and inference. Use pictures showing action that has already taken place and have students make observations and inferences about their observations. Comic strips, cartoons, coloring books, and comic books are good sources of pictures. Pictures of animals are also excellent for developing observation and inference skills. Organisms are adapted for their survival (e.g., coloration for protection, feet adapted for catching prey, feet for escaping predators, spines on cacti for protection, color to attract animals to eat the fruit like grapes, etc.)

2. *Mystery Boxes* are fun and intriguing as well as excellent activities for observations and inference making. Enclose unknown objects in a shoe box and have the students make as many observations as possible without opening the box. Try to involve as many senses as you can except sight, by providing the means to the student to feel or smell the object. Give students practice in accepting, modifying or rejecting inferences on the basis of additional information. Some objects that could be used include a sugar cube, bar of soap, a toothbrush, pine cone, popcorn, lemon, onion, or a stick of gum. There are many other objects that could be used.

3. *Unknown Gases* may be best done as demonstrations but students can participate. The following activities are wonderful for having students make inferences:

Oxygen

First you'll need:

- ✔ glass jar or clear plastic container
- ✔ enough 3% hydrogen peroxide to fill the jar about half full
- ✔ package of yeast
- ✔ candle (or splint) and matches

1. Fill the jar about half full with hydrogen peroxide and sprinkle some yeast into the peroxide. Have students record their observations. (What they *don't* see, smell, feel, or hear can be an observation too.)
2. Add a little more yeast, then light a candle or splint and lower the flame into the jar above the liquid. What happens?
3. Blow out the candle and lower the glowing wick into the jar. What happens? What inferences can be made from your observations? (This is really an oxygen generator but students can only *infer* the presence of oxygen as they observe the burning candle flame up and burn faster and the glowing wick burst into flame again.)

Carbon Dioxide

For this activity you'll need:

✔ baking soda
✔ vinegar
✔ candle
✔ matches
✔ small deli container or beaker
✔ peanut butter jar (or similar size jar)
✔ piece of clay

1. Place a few scoops of baking soda into the container.
2. Pour a similar amount vinegar into the soda in the container. What do you observe?
3. Lower a burning match into the container just above the solution. What happens? Do it again. (Students should see that the flame goes out. They might infer that carbon dioxide has been generated and that explains why the flame goes out.)
4. Secure the candle with the clay to the bottom of the inside of the peanut butter jar. Then light the candle.
5. Carefully *pour* the carbon dioxide gas over the burning candle. What happens? Why?

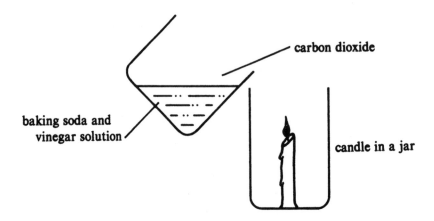

carbon dioxide

baking soda and
vinegar solution

candle in a jar

What are some possible reasons why that happened?

What are some other explanations for what you observed?

What might have caused that to occur?

What is a logical explanation for that?

In your own words explain why that happened.

How do you interpret this evidence?

What does this set of observations mean to you?

High Stakes Testing

A sample multiple-choice item from State Standardized Exams.

Which of these is most important for helping an animal move through soft sand or snow?

   f.   short thin legs
   g.   a long sturdy tail
   h.   a long neck
   i.   big wide feet ←

Virginia: SOL Grade 5 Spring 2001 Release Item.

www.humanities-interactive.org/texas/wtw/wtw_teacherlearning.htm
www.sasked.gov.sk.ca/docs/elemsci/g4fslc9.html

Technology Spotlight

Using Spreadsheets to Record, Display, and Interpret Data

Spreadsheets and databases are readily available and offer many opportunities for students to collect and interpret real data. Spreadsheet and database software are virtually found on all computers both in the school setting as well as on students' home computers. Although these applications are viewed by some as only personal productivity software used to keep records such as attendance and grade information, spreadsheets offer a powerful tool for teachers to provide regular opportunities for students to collect, display, and interpret real data.

Real data for a class or individual experiment is easily collected in a table utilizing a spreadsheet. The data from the spreadsheet can be translated into a graph with the touch of a button. This allows the instructional emphasis to be on selecting the most appropriate graph and then interpreting the graphed data rather than on the mechanics of drawing the graph. In the following example a teacher created a spreadsheet with an embedded graph. As students entered the results of their experiment, a bar graph displaying the amount of sugar in two brands of bubblegum was created at the same time the data was entered.

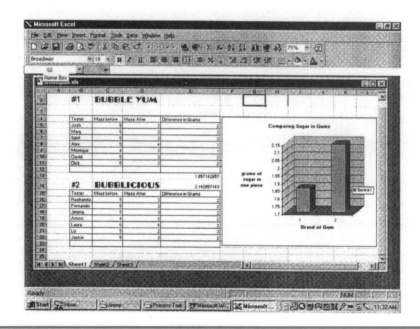

A Model for Assessing Student Learning

Assessment Type: Performance Task
Preparation

One way to assess a student's ability to construct inferences based on observations is to construct mystery boxes. Be sure to use identical boxes and objects when assessing children simultaneously. An example of a mystery box is one that uses two small objects that either roll or slide. Inexpensive and readily available rolling objects include BB's, ball bearings, plastic Easter eggs, and marbles. Metal washers and bottle caps and lids are good sliding objects.

Directions to the Student

1. Write your name on this line _____.

2. Check the box to be sure it is taped shut. Do not open the box.

3. The box contains one or more objects. Pick up the box and listen to the sounds as you gently shake it. Tilt the box and listen carefully.

4. Answer the following questions:
    _____ a. What shape is an object in the box?
    _____ b. What is one other property of an object in the box?
    _____ c. What is one kind of motion made by an object in the box?
    _____ d. Except for air, how many objects do you think are in the box?
    _____ e. Explain why you think your answer to question d is correct?

Scoring Procedure

1 point each for questions a, b, c, and e. Question d is not scored separately.

Acceptable responses for questions a, b, c, and e include:

a. flat, round, ball-shaped, like a coin, and so on.
b. hard, heavy, sounds like metal.
c. slides, rolls, glides, or drops.
e. student response has to support response to question d.

Chapter **6**

Predicting

National and State Standards Connections

- Develop predictions using evidence. (NSES 5–8)
- Develop and evaluate inferences and predictions that are based on data. (NCTM 3–5)
- Make predictions based on observed patterns and not random guessing. (California Content Standards: Grade 2)

MATERIALS NEEDED

To do the *Predicting* activities in this chapter you will need:

✔ A board at least 1 meter long and 30 cm wide to serve as a ramp for cans to roll down

✔ 2 unopened and un-dented cans of food (from the grocery store) to roll down the ramp. One can **must** contain only a very runny liquid—a can of beef or chicken broth works great. The second can **must** be the same height and have the same diameter as the first can. It **must** contain very chunky food. A can of diced tomatoes, thick salsa, or some type of chunky soup will work.

✔ A pencil, ruler or similar object to serve as a pendulum support

✔ A piece of string with an attached paper clip to serve as a hook

✔ A variety of small objects to serve as pendulum bobs (washers, sinkers)

✔ A meter stick

For the optional activity you will need:
✔ a glass jar (a liter or larger)
✔ enough particles to fill the jar (peas, marbles, rice, and so on)

103

"I think this end of the broom will weigh the same as that end of the broom," replied Hanna, a third grader, responding to the teacher's question.

Hanna's teacher had asked her to come to the front of the classroom for a science demonstration. The teacher handed Hanna a broom and told her to balance the broom horizontally on her finger. Hanna looked very puzzled but moved the broom handle carefully back and forth on her forefinger until she got it to balance. The broom, with its long wooden handle on one end and big clump of whisks at the other, teetered back and forth on Hanna's finger as the teacher marked the exact spot where Hanna's finger served as the fulcrum for the balanced broom.

The teacher asked the class to observe carefully, then asked, "If we take a saw and cut the broom into two pieces exactly on the mark where Hanna's finger is located, will the two pieces of broom weigh the same amount?"

Most students agreed that the two ends of the broom will weigh the same. Some others, though, disagreed and said the two ends of the broom will not weigh the same. The teacher asked several students to explain their predictions. As students explained *why* they predicted as they did, the teacher listened very carefully.

Predicting Is a Powerful Teaching and Learning Tool

A prediction is a forecast of what a future observation might be. In the scenario above, the teacher was able to expose beliefs the children already had about what it means when objects are balanced by asking the students to make predictions and to explain their predictions. Once students' current beliefs about a concept are revealed, a teacher can act to reinforce those that are correct, or confront those that are not correct and attempt to change them. In the case of the balancing broom, 'confronting' means the teacher had to be willing to take a saw and cut the broom so students could weigh both ends of the broom. If you are wondering about this yourself, balance a broom on your fingers. And if you are not quite sure both ends would weigh the same, get a saw, a scale of some type, and give it a try!

Predicting is closely related to observing, inferring, and classifying; predicting is an excellent example of one process skill being dependent on other process skills. The ability to construct reliable predictions depends on careful observations and inferences made about the relationships among observed events. When we observe similarities and differences, we use classification skills to impart order to those observations. Ordered observations often result in the recognition of patterns. We can then use these recognized patterns to predict what future observations might be.

When we hear our students ask questions such as, "If this happens, what will come next?" or "What will happen if I do this?" we know they are beginning to predict on their own.

———————————————————— Classroom Scenario ————————————————————

Goals

In these exercises you will learn to construct predictions based on patterns of observed evidence and to test your predictions for dependability. New observations can be used to revise both predictions and inferences.

Performance Objectives

After completing the activities in this chapter you should be able to:

1. Distinguish among observation, inference, and prediction.
2. Make predictions based on observed patterns of evidence.
3. Construct tests for predictions.
4. Use new observations to revise predictions and inferences.
5. Make predictions using graphed data.

Activity 6.1 Distinguishing among Observation, Inference, and Prediction

These brief definitions may help you distinguish among observation, inference, and prediction.

- Information gained through the senses: *Observation*
- Why it happened: *Inference*
- What I expect to observe in the future: *Prediction*

This activity is intended to give you practice in distinguishing among these important processes.

→ **GO TO** the supply area and obtain a ramp (a flat board or a shelf from a book case will do), a can of chicken broth and a can of diced tomatoes. Elevate one end of the ramp. Pick up a can of chicken broth and lay it on its side at the top of the ramp. Let it roll down the ramp. Do it again. Observe and record how the can behaves when it is rolling down the ramp:

_____

_____

_____

Hold the can of diced tomatoes in one hand and the can of chicken broth in the other. Predict what will happen if you place both cans at the top of the ramp and release them so they race one another down the ramp:

_____

_____

_____

Test your prediction by racing the two cans against one another. Which one won? _____

Try to explain *why* one can traveled down the ramp faster than the other (if it did).

_____

_____

_____

Brian, a fourth grade student, made the following statements after he raced his two cans down the ramp. Indicate whether each of Brian's statements is an observation, inference, or prediction.

1. "Both cans are the same size." _____.
2. "One can moves faster because of the way the stuff inside is arranged." _____
3. "I bet if I raced a can of chicken broth and a can of chicken noodle soup down the ramp, the can of noodle soup would win." _____

Self-Check

Compare your answers with someone else's or with those below.

1. observation (information gained through the senses)
2. inference (an explanation for the observation)
3. prediction (a forecast of what a future observation will be)

The process skills of observing, inferring, and predicting can be clearly defined and each is clearly distinguishable from the others. However, you will see later that there is also a great deal of interdependence among these processes.

We make sense of the world around us by observing things happen and then interpreting and explaining them. We often detect patterns in what we observe. When we think we can explain why things work the way they do, we construct mental models in our heads that at least temporarily serve to provide order to things. Often we use these mental models to predict occurrences that might happen in the future. Here are some examples of predictions:

- I see it is raining and the sun is coming out. There could be a rainbow.
- When I flip the switch, the lamp will light.
- The weak magnet picked up five paper clips; I predict the strong magnet will pick up more.
- If I release both balls at the same time, they will hit the ground at the same time.

Notice that each of the sample predictions is written in future tense. Each prediction statement is based on observations and patterns that have developed from past observations. How we explain and how we interpret what we observe affect how we predict.

A *map* of the process of predicting might look something like this:

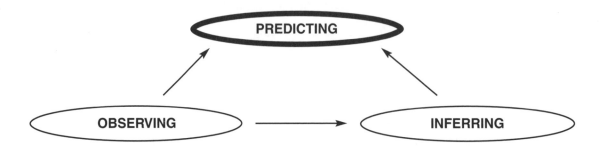

Predictions are *reasoned statements* based not only on what we observe but also on the mental models we have constructed to explain what we observe. Predictions are not just wild guesses because guessing is often based on little or no evidence.

In order to use the process skills of observing, inferring and predicting correctly you need to be able to clearly distinguish among them. The first activity was intended to provide you with some brief working definitions and to give you practice in distinguishing among observing, inferring and predicting. (You may want to refer to the separate learning activities for a thorough treatment of each process skill.)

Testing Our Mental Models

At one time in history many people believed that the Earth was flat. The Earth *looked* flat; that observation as well as others seemed to support the inference that the Earth was flat. People predicted that if sailors traveled far enough, their ship would fall off the Earth. Many people had a great deal of confidence in that prediction. Later, when sailors tested that prediction, they observed that their ship did *not* fall off the Earth. New observations caused people to change both their inference about how the Earth was shaped and their prediction about falling off the Earth.

Testing our predictions leads to making more observations that either support or do not support original predictions. When new observations are consistent with our predicted pattern of observations, we have even greater confidence in our prediction. However, when new observations do not support our original prediction, we may reject it and re-examine our observations. New observations lead to new inferences and new predictions. Therefore, our *map* of the process skill of predicting looks more like this:

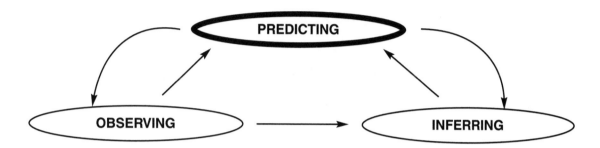

As new data (observations) are collected, theories (inferences) are proposed to explain what has been observed and to predict what has not yet been observed. In fact, for a theory to be accepted in science, it must meet a threefold test:

1. can explain what has already been observed
2. can predict what has not yet been observed
3. can be tested by further observation and modified as required by the new data.

Observing, inferring, and predicting are interconnected thinking skills. We use these skills to make sense of our world. Our ideas of how things work should be under constant review and subject to revision. Science should always be viewed as tentative; always subject to change as new observations result from testing our predictions.

Activity 6.2 Constructing Predictions Based on Observed Patterns

Predicting is stating what you think a future observation will be. You may recall that observations may be qualitative or quantitative in nature. While the activities you will do next involve both kinds of observations (qualitative and quantitative), you will be using some special tools to help manage the numbers in the quantitative observations you make.

Numbers are like any other bits and pieces of information; if you leave them lying around loose they have very little meaning. Charts and graphs help to organize information for easy review and retrieval. Organizing numbers in charts and graphs is also an effective way to communicate information to other people. Hopefully, you have noticed the use of several *concept maps* in this book that are intended to guide your thinking and to provide a framework into which you can classify new ideas. It may be helpful for you to think of charts and graphs as kinds of concept maps. They are particularly useful in showing the relationships between two or more ideas. This chart, for example, shows the relationship between the date and the time of day the sun appears to rise at a certain place on Earth.

Date	Sunrise Time	Date	Sunrise Time
January 1	7:24	May 1	5:00
January 15	7:20	June 1	4:31
February 1	7:12	July 1	4:33
February 15	6:52	August 1	4:56
March 1	6:35	September 1	5:25
March 15	6:08	October 1	5:54
April 1	5:42	November 1	6:28
April 15	5:21	December 1	7:01

It is important to help young students realize that *sunrise* is a misnomer because the sun does not really rise and set. It is the Earth turning on its axis that causes the sun to appear to rise and set.

But what about days in the year not shown in the table? Would it be possible to predict *sunrise* times for those days *not* directly observed? Let's see.

We make predictions by first looking for patterns. Answer the following questions designed to help you find a pattern in the observed *sunrise* times:

1. What time did the sun appear to rise on Jan. 1? _____ On Feb. 1? _____
2. Would you expect the *sunrise* time for Jan. 15 to be about halfway between *sunrise* times for Jan. 1 and Feb. 1? _____ Is it? _____ (Check the observed time.) (Note that Jan. 15 is not exactly *halfway* between Jan. 1 and Feb. 1 but it is close.)
3. Use the *halfway* method to predict the *sunrise* time for Feb. 15. (Try not to look until you have figured it.) What is your prediction? _____ Then check your prediction with the observed *sunrise* time in the table.
4. Try this one. Predict the sunrise time for October 15. _____ Then check the answer in the Self-Check.
5. Suppose you wanted to predict a sunrise time for a date that was not halfway between two other given dates? Using the table, determine the sunrise time for September 10 and then check your answer in the Self-Check.

Self-Check _____ Activity 6.2

For #4. If you used the *halfway* method to find the predicted sunrise time for October 15, your calculations probably look something like this:

```
November 1    6:28
– October 1   5:54
              34  minutes difference
```
Half of 34 minutes is 17 minutes.

```
October 1    5:54
+            17  minutes
             6:11  predicted sunrise time for October 15
```

Activity 6.3 **Practice in Making Predictions**

When making predictions it is important to:

1. Collect data through the careful use of your senses (observing)
2. Search for patterns of characteristics and behaviors (classifying)
3. Formulate cause and effect relations (inferring)
4. Construct statements about what you think future observations will be (predicting)
5. Test the dependability of the prediction.

This activity involves the use of pendulums.

→**GO TO** the supply area and pick up a piece of string (about 50 cm or so), a paperclip to act as a hook, a selection of smaller and larger objects to act as 'pendulum bobs,' such as washers and fishing sinkers, a meter stick, and materials to make a pendulum support, such as a pencil or ruler and some tape. You will also need a watch or clock with a second hand.

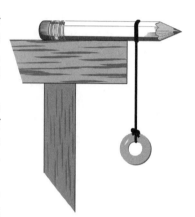

Tape a pencil or similar object to a desk or table so that it overhangs the edge. Tie a paperclip to one end of a piece of string. Tape the other end of string to the pendulum support you made. Make a hook out of the paperclip and attach one of the pendulum bobs. By taping the string and using a paperclip hook it will be easier for you to change the string length and the bobs.

Set the pendulum in motion and observe several full swings of the pendulum. A pendulum completes a full swing when it makes a *roundtrip* back to its original position.

Before you begin exploring with these materials, predict which of the following variables might affect the time it takes for a pendulum to make a full swing:

	Yes	No	
1.	_____	_____	the diameter of the bob
2.	_____	_____	the mass of the bob
3.	_____	_____	the location of where the pendulum is released
4.	_____	_____	the length of the pendulum
5.	_____	_____	the type of path you create (circular or back and forth)

In your explorations, find out what affects the length of time it takes for the pendulum to make a full-swing (roundtrip).

Compare your answers with someone else's or refer to the following Self-Check.

Self-Check _____ Activity 6.3

You probably found very noticeable differences in the roundtrip time of the pendulum when you changed the length of the pendulum.

Activity 6.4   Making Dependable Predictions

While you were exploring with pendulums, you were making observations and gathering data about pendulums and motion. Through exploration and observation you began making inferences as to what might affect round-trip time for the pendulum. Discovering *patterns* will enable you to make dependable predictions about the behavior of pendulums. By improving your inferences about what affects a pendulum's swing, you increase the likelihood that your predictions are correct. In other words, you are building *confidence* in your predictions.

You may also increase the amount of confidence you have in a prediction by arriving at the same predicted values by different methods. The closer the agreement between predicted values arrived at by different methods, the greater the confidence you may have in the prediction.

In this activity you will observe what happens when the length of a pendulum is systematically changed. Your observations will then be a basis for making predictions about the motion of the pendulum. Follow directions carefully.

Step 1: In this part of the activity you will be working with columns 1 and 2 of the table that follows. Adjust the length of the pendulum to 15 cm (measure to the end of the bob). Count the number of full swings (round trip) in a 30 second interval. Record the number of full swings in the blank space in column 2.

Repeat the procedure for 25 cm, 35 cm, and 45 cm. These are the observed number of full swings for those pendulum lengths.

How Does Varying the Length of a Pendulum Affect Its Motion?

1 Length of the Pendulum (cm)	2 Number of Observed Swings	3 Predicted from Observing Patterns	4 Predicted from Graphing the Data	5 Number of Observed Swings
15				
20				
25				
30				
35				
40				
45				

Step 2: Examine the data in columns 1 and 2. Without swinging the pendulum, predict the number of swings the pendulum would make in a 30 second interval for 20, 30, and 40 cm. Enter these predictions in the blank spaces in column 3. These represent your predictions using the method of observing patterns.

There is more than one way to make predictions using this data. A second method involves graphing the data.

Step 3: Using the data from column 2, plot the values for 15, 25, 35, and 45 cm. On the following graph, draw a smooth curve through the plotted points.

How Does Varying the Length of a Pendulum Affect Its Motion?

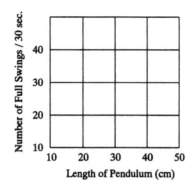

Compare your graph with the one below.

Self-Check _____ Activity 6.4 ✔

Your graph probably looks something like this:

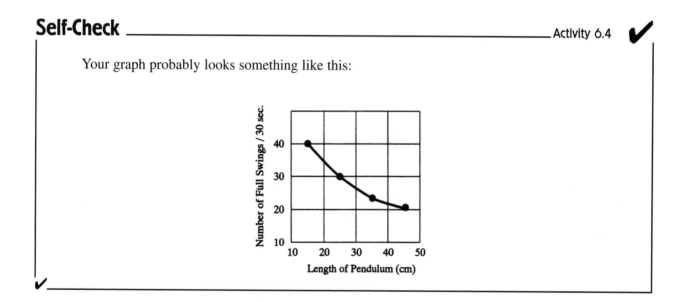

Step 4: By reading the graph, predict the number of swings for 20, 30, and 40 cm. Enter these predictions in column 4 of the table. You made these predictions by the method of graphing data. Predictions made between observed data points are called *interpolations.* Predictions made beyond the observed data re called *extrapolations.*

Compare the values you predicted using the two methods of observing patterns and graphing the data. How similar were they? The closer the agreement between the predicted values obtained by different methods, the greater the confidence you should have in your prediction. How confident are you that your predicted values are correct?

Step 5: Put the predictions to the test by observing the number of full swings of the pendulum at 20, 30, and 40 cm. Enter these numbers in the table under column 5. These are your observed values. Compare these to your predicted values. They should be fairly close.

Self-Check _____ Activity 6.4 ✔

Your table may look something like this:

How Does Varying the Length of a Pendulum Affect Its Motion?

1 Length of the Pendulum (cm)	2 Number of Observed Swings	3 Predicted from Observing Patterns	4 Predicted from Graphing the Data	5 Number of Observed Swings
15	40			
20		35	35	34
25	30			
30		27	27	27
35	25			
40		23	23	22
45	21			

✔

Activity 6.5   **Prediction and Chance**

Few things are as dependable as sunrise time and the swing of a pendulum. To illustrate this point, do the following activity. You will need access to a supply of students from your class and at least one other class (the closer to a total of 50 students, the better). These students will be your *population.*

Examine the earlobes of several people seated around you to help you determine what an attached earlobe looks like and what a free (or unattached) earlobe looks like. Your instructor can help you make this distinction. In the boxes below, do your best to make a drawing of each type of earlobe:

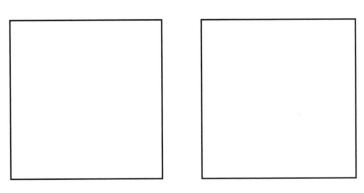

attached earlobe   **free earlobe**

Observe 5 students seated near you and tally the number of students who have attached earlobes and the number of students who have free earlobes. Label this group of students the *small* sample. Enter your tallies in the following table as you collect them.

Observed and Predicted Number of Students with Attached and Free Earlobes Using a Small Sample

Observed number of students with *attached* earlobes in the *small* sample (5 students)	*Observed* number of students with *free* earlobes in the *small* sample (5 students)	*Predicted* number of students with *attached* earlobes in the population	Predicted number of students with *free* earlobes in the population

Using your small sample tallies and your math skills, predict the number of students in the total population (all 50 or so students) who will have attached earlobes and free earlobes. Enter your predictions in the table above.

Now increase the size of the sample from 5 to 20 students. Call this the *larger* sample. It should include the original 5 students surveyed and 15 additional students. Enter your tallies for this larger sample in the following table:

Observed and Predicted Number of Students with Attached and Free Earlobes Using a Larger Sample

Observed number of students with *attached* earlobes in the *larger* sample (20 students)	*Observed* number of students with *free* earlobes in the *larger* sample (20 students)	*Predicted* number of students with *attached* earlobes in the total population	*Predicted* number of students with *free* earlobes in the total population

Using your *larger* sample tallies, predict the number of students in the total population (all 50 or so students) who will have attached earlobes and free earlobes.

How good are the predictions you made? Is the small sample just as good a predictor as the larger one? You simply do not know the answers to these questions unless you can compare your predictions with the actual results from the total population. Fortunately, in this activity you can easily determine the total population results and compare them to your predictions. Because you have already observed 20 students, you will need to add 30 more to this number to complete the total population. Do that now and record your tallies in the following table:

Observed Number of Students with Attached and Free Earlobes in the Total Population

Observed number of students with attached earlobes in the total population	Observed number of students with free earlobes in the total population

Compare the predictions you made using the two sample sizes with the actual population results.
Answer these questions:

1. Did sampling serve as a good predictor of total population numbers? _____

   _____

2. Did the small sample serve as just as good a predictor as the larger one? _____

   _____

3. What factors may have affected the accuracy of your predictions? _____

   _____

 **Optional Activity   Estimating Large Quantities**

Have you ever wondered . . .

- How many blades of grass are in a lawn?
- How many words are in a book?
- How many grains of sand are in a beach?
- How many leaves are in a tree?
- How do they know how many eagles are alive?
- How many words are in a newspaper?
- How many stars are in the sky?
- How many grains of rice are in a bag?
- How many flakes are in a box of cereal?
- How many jelly beans are in that jar so you could win a bike?

As you can see, sometimes it is more practical to estimate than actually count. As a test for your ingenuity,
design and conduct a method to determine the number of items in a container. The container could be a container of marbles, buttons, rice, or anything else your instructor is devious enough to contrive.

Compare your method with those in the Self-Check, your instructor's, or someone else's.

Self-Check _____ Optional

Listed below are some methods that may be used to predict large numbers of things. If you discover other methods, add them to the list.

1. **Sampling by volume:** Count the number of items in a small sample-size volume, such as a
   3 oz. size paper cup. Determine the number of the smaller volumes in the total volume and multiply the result by the number of counted items in the sample.
2. **Sampling by mass:** Find the mass of a small, easy to count quantity as well as the mass of the
   total quantity. Divide the mass of the total quantity by the mass of the sample. Multiply the
   result by the number of counted items in the sample.
3. **Sampling by area:** Spread the items one layer thick as evenly as possible over a grid-lined
   paper, such as graphing paper. Count the number of items in a small square section of the grid.
   Multiply the result by the number of small square sections that are in the total covered grid.

4. **Halving and doubling:** Divide the total quantity into two parts. Continue halving each part until you have a small enough quantity to count. Record the number of times you halved the items. Now reverse the process. Double the number of items you counted and then double this amount. Continue doubling the new amount until you have doubled as many times as you originally halved the items.

 # Self-Assessment

Predicting

I. Here is a graph showing the mean (average) monthly high temperatures for a particular city. Examine the graph and answer the questions that follow.

1. What basic science process skill was used to gather the information for the graph? _____
2. What basic science process skill was used to organize the data? _____
3. What basic science process skill would you use to explain why the curve on the graph has this particular shape? _____
4. What was the mean high temperature recorded for the month of February? _____ April? _____
5. What basic science process skill would you use to forecast the mean high temperature for the month of May? _____
6. Predict the mean high temperature for the month of March. _____
7. In what part of the United States would you predict finding a city with a similar yearly temperature graph? _____
8. How does your confidence in your prediction for question #6 compare with your confidence in your prediction for item #7? _____

II. 1. Keisha had a jar containing 100 pennies. (Twenty-five of the pennies were dated 2000, 50 of the pennies were dated 2001, and the remaining 25 pennies were dated 2002, but Keisha did not know this.) Without looking, Keisha took a sample of ten pennies from the jar and examined their dates. How many pennies would you predict she found for each of the years?
Year 2000 pennies = _____
Year 2001 pennies = _____
Year 2002 pennies = _____
2. When Keisha looked at the pennies, she found that five pennies were dated 2000, four pennies were dated 2001, and one penny was dated 2002. Keisha's results were probably not similar to your predictions. As her teacher, what two suggestions might you make to Keisha for her to obtain a better idea of the true distribution of pennies in the jar?

_____

_____

Compare your answers with the Self-Assessment Answers.

Self-Assessment Answers

I.   1. Observation
     2. Classification
     3. Inference
     4. about 2 °C about 14 °C
     5. Prediction
     6. about 8 °C
     7. The city was actually Detroit, Michigan
     8. You should have felt much more confident in the predictions you made for item #6 than the prediction you made for item #7. The difference lies in that your prediction for #6 was based upon careful and comprehensive observations and a definite pattern in the data. Your prediction for item #7 was most likely a guess.

II. 1. Based on the date percentages of the pennies (25%, 50%, and 25%), a reasonable prediction for a sample size of ten would be two or three dated 2000, five dated 2001, and two or three dated 2002.
     2. Keisha could take a larger sample, or take several small samples and average the results.

Technology Spotlight

Using and Creating WebQuests

A WebQuest is an inquiry-oriented task that students solve independently or cooperatively using information from the Internet. WebQuests all share the same basic elements. These include an introduction, task, information resources, processes, learning advice, and evaluation. The best examples are those that take students beyond just retelling and summarizing information to engaging them in problem solving, creativity, design, and judgment. Structured templates are available to help first time users at *http://webquest.sdsu.edu/ LessonTemplate.html.*

WebQuests are designed to focus on using information rather than looking for it. Originally developed in 1995 at San Diego State University, there are now literally thousands of WebQuests on the Internet covering all grades and subject matter. Start by exploring the WebQuests others have created. You may find a WebQuest that fits your needs or only needs minor modification for your purposes. Try locating suitable WebQuests using a search engine. Use quotation marks to narrow the search to a reasonable number of sites by enclosing the word WebQuest and your topic of interest, such as "moon WebQuest."

Another good place to start is at *http://webquest.sdsu.edu/materials.htm* to get an overview of WebQuests and links to related articles and other helpful resources. At the time of this writing an excellent list of WebQuests organized by subject and grade could be found at *http://sesd.sk.ca/teacherresource/webquest/webquest.htm.*

IDEAS FOR YOUR CLASSROOM

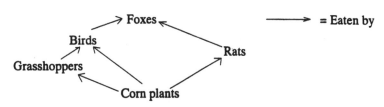

1. *Food Webs.* Here is a simple food web. Predict what would happen if any one organism were removed from the web.

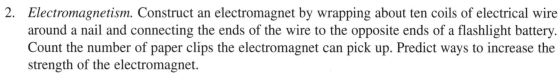

$\longrightarrow$ = Eaten by

2. *Electromagnetism.* Construct an electromagnet by wrapping about ten coils of electrical wire around a nail and connecting the ends of the wire to the opposite ends of a flashlight battery. Count the number of paper clips the electromagnet can pick up. Predict ways to increase the strength of the electromagnet.

3. *Inclined Planes.* Make an incline by placing a book under one end of a ruler. Place a marble in the groove of the ruler at the top of the incline. Release the marble and measure the distance the marble travels. Predict how the distance the marble travels can be increased.

4. *Electrical Conductors and Insulators.* Construct an electrical circuit by connecting the opposite ends of a flashlight battery to a flashlight bulb using two wires. Be sure the circuit works (the bulb lights). Test various objects to see if they are good conductors of electricity by first making a gap in the circuit between the bulb and one wire and inserting the object to be tested in the gap. If the bulb lights, the object is a conductor. With a new set of objects made of various materials (glass, rubber, plastic, wood, different kinds of metals) predict which ones are conductors.

5. *Sound.* Add different amounts of water to several soda bottles and predict which will produce the highest pitch when struck sharply with a pencil. Which will give the highest pitch when someone blows into the top of the bottle?

6. *Combustion.* Invert jars of varying volumes over burning birthday candles. Predict which candle will burn the longest.

7. *Sound.* Hammer nails in pairs at various distances apart on a board. Stretch a rubber band over each pair of nails. Predict which rubber band will make the highest pitch when plucked. Predict a way to change the pattern.

What do you think will happen if . . . ?

Predict what might happen next.

If this is changed, what will happen to that?

How will changing this variable affect that variable?

What do you think might happen next?

High Stakes Testing

A sample multiple-choice item from State Standardized Exams.

What will probably happen if the plant is turned around so that it faces the opposite direction?
  A.   It will stand up straight.
  B.   Its leaves will fall off.
  C.   It will bend back toward the sunlight. ←
  D.   Its leaves will turn yellow.

Illinois: ISAT Grade 4 Sample Test.

http://illuminations.nctm.org/lessonplans/3-5/coasters/index.html
www.askeric.org/Virtual/Lessons/Science/Process_Skills/SPS0008.html
www.sasked.gov.sk.ca/docs/elemsci/g4fslc10.html

WEBSITES FOR ACTIVITIES

A Model for Assessing Student Learning

Assessment Type: Performance Task[1]

Teacher Preparation

Station Two—Water on Objects

Materials (per station)
- ✔ five 5-cm square pieces of white or tan paper napkins for each class
- ✔ five 5-cm square pieces of buff manila folder for each class
- ✔ five 5-cm square pieces of white or tan paper towel for each class
- ✔ one 5-cm square piece of white, unlined index card
- ✔ one clear plastic sealable sandwich bag
- ✔ one eye dropper
- ✔ one hand lens
- ✔ one small container (100 mL)
- ✔ one large container (150–250 mL)
- ✔ paper towels
- ✔ one non-water-soluble ink marker or stamp pad
- ✔ and a direction sheet for Station Two

Preparation

1. Label each 5-cm square piece of white or tan paper napkin "A."
2. Label each 5-cm square piece of buff manila folder "B."
3. Label each 5-cm square piece of white or tan paper towel "C."
4. Label the 5-cm square piece of white, unlined index card "X."
5. Place "X" in the plastic sandwich bag and seal it.
6. Label the bag "Do Not Open."
7. Label the 100-mL container "Water," and fill with 50 mL of water.
8. Label the large container "Waste."
9. Tape the direction sheet to the lower left side of the tabletop.
10. Place all materials on the tabletop as shown.
11. Make sure that fresh sets of papers A, B, and C are available for every student who will be tested at the station.

Directions to the Students

1. Check the materials
   - ✔ Three piece of paper (marked A, B, and C)
   - ✔ Plastic bag containing a paper marked X
   - ✔ Hand lens
   - ✔ Dropper
   - ✔ Container of water
   - ✔ Container marked "Waste"
2. Read question 1 on the answer sheet for **Station Two.**
3. Use the pieces of paper marked A, B, and C only. **(Do not open the plastic bag.)**
4. Place **one** drop of water from the container on each piece of paper. Use the hand lens to look at what happened to each drop of water.
5. Answer question 1 on your answer sheet.
6. Look at the paper marked X inside the bag. **Do not open the bag!** You may look at paper X with the hand lens.
7. Answer questions 2 and 3 on your answer sheet.
8. Put the used pieces of paper marked A, B, and C in the container marked "waste."

Answer Sheet

1. What happened to the drop of water on each piece of paper?
   On paper A, the drop of water _____
   On paper B, the drop of water _____
   On paper C, the drop of water _____
2. You can not put water on Paper X. But **if** you could, **predict** what would happen to the drop of water. _____
3. Why did you predict this would happen? _____

Scoring Procedure

For question 1: Maximum score is 3 points

Sample of Acceptable Answers:
- On papers A and C, the drop of water was absorbed; soaked in; spread out; got bigger; expands; fills up/makes squares or blocks; goes through.
- On paper B, the drop of water was not absorbed; sits on top; stays in ball; stays the same; doesn't spread: stays a drop; won't go through; bubbles on top.

For question 2: Because paper X is similar to paper B, the student prediction must indicate that he or she has observed the similarity of the two papers.

Sample of Acceptable Answers:
- the same thing that happened to paper B; it would not be absorbed; the water would sit on top.

For question 3: The students explain the prediction made for question 2, such as:
- the paper is hard; paper X is shiny; it has no holes; looks like paper B.

[1]Doran, R.L., Reynolds, D., Camplin, J., and Hejaily, N. (1992). Evaluating Elementary Science. *Science and Children,* 30(3), 33–35, 63–64.

Selected Resources for Teaching Elementary and Middle School Science

Activities Integrating Math and Science
Grades K–8
P.O. Box 8120
Fresno, CA 93747
http://www.aimsedu.org/documents/puznac.html
This website has an online catalog, teacher resources, professional development information, state correlation documents, puzzles and activities.

Bottle Biology
Grades K–6
Kendall/Hunt Publishing Company
4050 Westmark Drive
Dubuque, IA 52004, 800/258-5622
http://www.sci.mus.mn.us/sln/tf/books/bottlebiology.html
This website has ordering information.

Full Option Science System (FOSS)
Grades K–6
Delta Education, Inc.
P.O. Box 915
Nashua, NH 03063
http://www.delta-education.com

Great Explorations in Math and Science (GEMS)
Lawrence Hall of Science
University of California
Berkeley, CA 94720
http://www.lhs.berkeley.edu/gems/
This website has information about GEMS teacher guides and handbooks, online activities, professional development opportunities.

Lab-Aids
kit materials for elementary, middle and high school
Lawrence Hall of Science
University of California
Berkeley, CA 94720, 510/642-7771
http://www.Lab-Aids.com
This website has information about their applied science concept kits, teacher guides, student worksheets, and National Science Education Standards.

National Science Teachers Association
Journals available with membership: *The Science Teacher, Science Scope, Science and Children*
1840 Wilson Blvd.
Arlington, VA 22201-3000
http://www.NSTA.org/
This website also has teacher resources including journals and books, teachers' grab bag, supplier guide for members and nonmembers, links to online National Science Education Standards and recommended websites.

Science and Technology for Children (STC)
Grades 1–6
Carolina Biological Supply Company
2700 York Road
Burlington, NC 27215, 800/334-5221
http://www.carolina.com/
Website also has a product catalog, links to teacher resources, and "ask a professor" service.

Science Activities for the Visually Impaired & Science Enrichment for Learners with Physical Handicaps (SAVI/SEPH)
Grades 3–7
Center for Multisensory Learning
Lawrence Hall of Science
University of California at Berkeley
Berkeley, CA 94720
http://store.yahoo.com/lawrencehallofscience/savscienacfo.html
This website has information about available print, video and kit materials.

Science Education for Public Understanding (SEPUP)
modular materials (4–6, 6–9)
Lawrence Hall of Science
University of California
Berkeley, CA 94720, 510/642-7771
http://www.sepup.com
This website has information about their issue-oriented science modules, teacher resources, professional development opportunities, a link to the National Science Education Standards and other technology links.

Wisconsin Fast Plants
Grades K–12
Kendall/Hunt Publishing Company
4050 Westmark Drive
Dubuque, IA 52004, 800/258-5622
http://www.fastplants.org/
This website has information about ordering and growing Fast Plants, activities and resources.

Decision Making 1

Now that you have learned the Basic Science Process Skills, you can use what you have learned to improve existing science curricula. In learning the science process skills, you not only mastered the skills, but you also learned something about how these skills can be learned. By using this knowledge you can begin making some important instructional decisions about teaching science, especially the science process skills. In this section you will focus on the application of what you know about the science process skills to improve elementary school science textbook activities and other science curricula materials. The decisions you make can significantly enhance the quality of science in which your students are engaged.

Read *Typical Textbook Activity Example A* on the next page. Think about how you might change the activity and the suggested teaching strategies to better emphasize the science process skills.

As you study the sample activity, look at both the content and skills your students will be learning and how they would be learning them.

Ask yourself, *How will I provide opportunities for my students to:*

- *use their senses?*
- *classify and form concepts?*
- *measure and quantify their descriptions of objects and events?*
- *communicate orally and in writing what they know and are able to do?*
- *infer explanations and change inferences as new information becomes available?*
- *predict possible outcomes before they actually observe?*

On a separate piece of paper, with these questions in mind, write what you consider to be strengths and weaknesses of *Example A*. Then consider how you might change the activity to improve the weak areas.

After you have studied *Example A* and thought about how you might change it, turn the page and look at our modified version of *Example A*. The changes made to *Example A* are only a few modifications that could be made to this activity to better emphasize the process skills. Your ideas for modifying this activity may have been different and even better.

Typical Textbook Activity Example A

ACTIVITY

> **Science Skills**
>
> observing, collecting data, making a graph

Fruits and Seeds

How many seeds are in some fruits?

1. Remove all the seeds from one fruit and place them on a paper towel.
2. Count and record the number of seeds.
3. Do steps l and 2 for each fruit.
4. Make a graph to show the number of seeds in each fruit.

Which fruit has the most seeds?

Which fruit has the least seeds?

TO THE TEACHER

> **Time:** 30 min.
>
> **Groups:** 4 students per group
>
> **Materials:**
> - ✓ 4 different fruits per group
> - ✓ spoons
> - ✓ paper towels
> - ✓ graph paper
>
> **Objective:** Students will count the seeds in a variety of fruits and graph the results.

Lesson Setup

- Use a variety of fruits, some with many seeds and some with just one.
- Cut the fruit for the students.
- Every student should make a graph.

Suggestions

- Students should use the spoons to remove the seeds.
- Remind students to keep fruits and seeds on the paper towels.
- Show students examples of graphs.
- Check to see that students are constructing and labeling graphs accurately.

Modified Textbook Activity Example A

Here are some modifications to this activity that *we* made.
Your modifications may be even better.

use 6 fruits
- *large grapefruit*   *-peach*
- *apple*   *-orange*
- *cucumber*   *-muskmellon*

ACTIVITY

Science Skills
communicating, classifying, observing, collecting data, making a graph, *predicting*

1. *Have the students name the fruits and classify them according to size.*

2. *Have students find other ways to classify them.*

3. *Have students predict how many seeds are in each and record their predictions on a piece of paper. Help them design a chart.*

Fruits and Seeds

How many seeds are in some fruits?

4. ~~1.~~ Remove all the seeds from one fruit and place them on a paper towel.

5. ~~2.~~ Count and record the number of seeds. *Compare with their predictions.*

6. ~~3.~~ Do steps 1 and 2 for each fruit.

7. ~~4.~~ Make a graph to show the number of seeds in each fruit.

8. *As a class, make a large graph compiling the data from each group.*
Which fruit has the most seeds?

Which fruit has the least seeds?

9. *Ask each student to write what he or she found out using a word bank on the board.*

TO THE TEACHER

Time: ~~30~~ min. *40*

Groups: 4 students per group

Materials:
- ✓ ~~4~~ *6* different fruits per group
- ✓ spoons
- ✓ paper towels
- ✓ graph paper

Objective: Students will count the seeds in a variety of fruits and graph the results.

Lesson Setup

- Use a variety of fruits, some with many seeds and some with just one.
- Cut the fruit for the students.
- Every student should make a graph.

Suggestions

- Students should use the spoons to remove the seeds.
- Remind students to keep fruits and seeds on the paper towels.
- Show students examples of graphs.
- Check to see that students are constructing and labeling graphs accurately.

Fruit	Number of Seeds	
	Predict	Count
~~		
~~		
~~		
~~		
~~		

Safety tips: Remind students not to eat the fruit. Ask students to wash their hands when finished.

Here is another textbook activity example. Your task is to modify this activity to emphasize the process skills as modeled in the previous example. It may help you to review the questions on page 123 and to note the strengths and weaknesses of this activity. Then make your changes right on this activity page. When you are done, see the page following for modifications we made.

Typical Textbook Activity Example B

ACTIVITY

In What Order Do the Parts of a Young Plant Develop?

You Will Need:
Seeds: corn and bean (vine beans work best), soaked in water overnight; paper towels, staples, water, plastic sandwich bags

Follow This Procedure:

Use your thumbnail to split open one seed and observe the inside.

food for plant
seed coat
embryo plant

Make a seed germinator by following these directions:

1. Fold a paper towel and slip it into the plastic sandwich bag to line the bag.
2. Make a row of staples about 4 cm from the bottom of the bag to form a shelf on which the seeds will sit.
3. Place about five seeds inside the bag just above the row of staples.
4. Holding the bag upright, slowly add water to moisten the paper towel and allow a little water to accumulate below the staples. It will be important to keep the towel moist all during germination.
5. Close the bag, and tack it to the bulletin board where it can be easily observed.

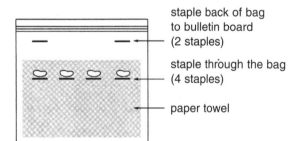
staple back of bag to bulletin board (2 staples)

staple through the bag (4 staples)

paper towel

The seeds should begin to sprout within a few days. Record observations of the seed each day.

In what order do you see the parts of the young plant develop?

Modified Textbook Activity Example B

Here are some modifications to this activity that *we* made.
Your modifications may be even better.

ACTIVITY

In What Order Do the Parts of a Young Plant Develop?

Follow This Procedure:

Use your thumbnail to split open one seed and observe the inside.

Have students make a drawing of their seed, list observations and compare with other seeds.

Make a seed germinator by following these directions:

1. Fold a paper towel and slip it into the plastic sandwich bag to line the bag.
2. Make a row of staples about 4 cm from the bottom of the bag to form a shelf on which the seeds will sit.
3. Place about five seeds inside the bag just above the row of staples.
4. Holding the bag upright, slowly add water to moisten the paper towel and allow a little water to accumulate below the staples. It will be important to keep the towel moist all during germination.
5. Close the bag, and tack it to the bulletin board where it can be easily observed.

Have students use hand lenses to make detailed observations; record detailed observations. Ask students to predict one week's growth and direction of growth.

The seeds should begin to sprout within a few days. Record observations of the seed each day.

In what order do you see the parts of the young plant develop?

Have students organize data in chart or graph form. Ask students to predict what would happen if seeds were placed in darkness, soapy water, oil, and so on.

You Will Need:

Seeds: corn and bean (vine beans work best), soaked in water overnight; paper towels, staples, water, plastic sandwich bags

Have students keep a journal in which they describe plant growth. Include regular entries of height of plant, number of leaves, diameter of stem and so on.

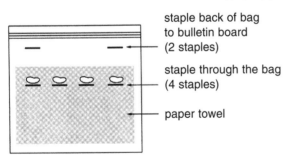

staple back of bag to bulletin board (2 staples)

staple through the bag (4 staples)

paper towel

Introduce the idea of quantitative and qualitative observations. Ask students to classify as quant. or qual.

Modifying Real Textbook Activities

The textbook examples you have just studied are typical of the kinds of science activities you might find in an elementary school textbook and other science curricula materials. To gain a little more experience at modifying materials to emphasize the science process skills and to become acquainted with real textbook activities, you have one more task to complete.

Obtain an elementary science textbook or other curricula materials for grade K, 1, 2, or 3. Locate an activity and modify it to better emphasize the basic science process skills just as you did in the previous examples. You might refer again to the questions on page 123 that focus on these skills. You may find it helpful to think of this task as a three-step procedure:

1. Examine the activity.
2. Identify parts that could be improved.
3. Add your improvements.

For feedback on your modifications, see your instructor, or try your modified activity with children and assess their skills. When you have completed this task, you will be ready to learn the Integrated Science Process Skills in Part Two of the book.

Part 2

Integrated Science Process Skills

Now that you have mastered the basic science process skills, you are ready to learn the skills that lead to *experimenting.* By combining these new skills with the basic science process skills, you can create a classroom climate where children explore, investigate, and discover.

In classrooms where students are learning the integrated science process skills, they inquire about how things work and they seek answers to their own questions by designing and conducting experiments. Rather than relying on their teacher and textbook to supply answers, these children ask themselves, *How can I find out?*

The integrated science process skills include identifying variables, constructing hypotheses, analyzing investigations, tabulating and graphing data, defining variables, designing investigations, and experimenting. Learning these skills empowers students to answer many of their own questions. Students who have learned the integrated skills have the tools to interpret what they observe and to design investigations to test their ideas.

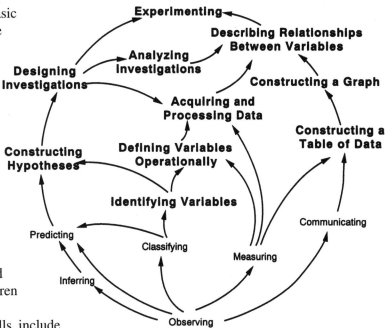

The integrated skills are not separate and distinct from the basic skills. The basic skills provide the foundation for the more complex integrated skills. For example, the predicting skills you learned in Part One are used to construct hypotheses. In Part Two you will learn that a hypothesis is a special kind of prediction that sets the stage for investigating relationships in science. You will find, as your students also will, that experimenting leads to asking more questions and conducting more experiments. Experimenting is a form of problem solving that requires the integration of all the other thinking skills.

Part Two of this book begins with the integrated science process skill of *Identifying Variables*. Each time you learn a new skill, ask yourself the same questions you asked in Part One:

Teaching Children

How am I learning this skill?
How will I teach this skill to students?

After you have completed the Integrated Science Process Skills, you will be asked to make some instructional decisions about how you might teach these same skills to your students.

Chapter 7

Identifying Variables

- Design and conduct a scientific investigation. . . . identify and control variables. (NSES 5–8)
- Analyze change in various contexts. (NCTM 3–5)
- Recognize well-designed procedures that identify important variables within an investigation. (Maryland Learning Outcomes: Grades 4–5)

MATERIALS NEEDED

- ✔ safety goggles
- ✔ 4 small identical containers (beakers or plastic cups)
- ✔ Celsius thermometer
- ✔ plastic spoon to use as a scoop
- ✔ large container (about a liter for holding water)
- ✔ calcium chloride (chemical used to control ice on roads)
- ✔ graduated cylinder

Purpose

In this chapter, you will be learning one of the skills needed when conducting an investigation. This important skill of identifying variables will be used throughout this section whenever you analyze how someone else conducted an investigation or whenever you plan and carry out an investigation of your own.

Objectives

After studying this chapter you should be able to:

1. identify the variables in a written statement or description of an investigation.
2. classify the variables as independent (manipulated) or dependent (responding).

Sometimes the most amazing things happen with the simplest of materials. In Chapter 1 you dissolved four different colored Gobstoppers in room temperature water. Gobstoppers are a kind of hard candy similar to 'jawbreakers.' As you observed them dissolving over time, you made some surprising observations.

If you have not yet conducted this activity, fill a pint-size deli container or similar container with enough room temperature water that will cover a Gobstopper, about 2 cm deep. Evenly place four different colored Gobstoppers around the container and sit back and watch as they dissolve. Note how the colors spread out through the water; puzzle over the clear areas that appear between colors as each Gobstopper goes through a series of color changes.

Dissolving takes a little time; could you speed up the dissolving time by using warmer water? What other things could you change that might affect how Gobstoppers dissolve? Here is a systematic way to brainstorm a list of possible factors that could be changed. **How could the set of Gobstopper materials (Gobstoppers, water, and container) be changed to affect the way Gobstoppers dissolve?**

Gobstoppers	**Water**	*Container*
Brand	Temperature	Shape
Quantity	Amount	Type of bottom
Color	Type	Material
Temperature	pH	Depth
Diameter	Carbonated or not	Diameter

These are just some of ways these materials could vary that might affect the way Gobstoppers would dissolve. Is there one of these factors that makes you wonder if changing it would make any difference?

The best way to become comfortable with science is to do science. You need to investigate a bit, use some equipment, and get your hands involved. To accomplish this and also learn how to identify variables, carry out the following activities.

➡**GO TO** the supply area and locate the following items:

- ✔ safety goggles
- ✔ 4 small identical containers (beakers or plastic cups)
- ✔ Celsius thermometer
- ✔ plastic spoon to use as a scoop
- ✔ large container (about a liter for holding water)
- ✔ calcium chloride (chemical used to control ice on roads)
- ✔ graduated cylinder

Activity 7.1

1. Allow a large container of tap water to reach room temperature. Use a graduated cylinder to fill each container with 75 mL of water.

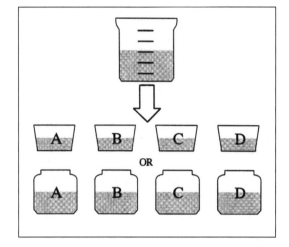

2. Measure the initial temperature of the water in one of the containers (call it container A). Put on your safety goggles and add 1 level scoop of calcium chloride and stir it until it dissolves.

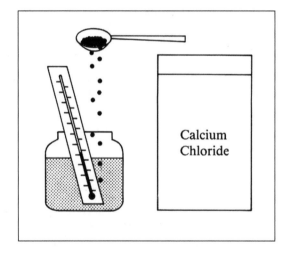

Measure the temperature of the water as soon as the calcium chloride dissolves. Subtract the initial temperature from the final temperature of the water.

a. What was the initial temperature of the water **before** adding the calcium chloride? _____
b. What was the temperature of the water **after** adding the calcium chloride? _____
c. What happened to the temperature of the water in the container? _____
d. How many degrees did the temperature change when you added one level scoop of calcium chloride? _____

You probably found that the temperature increased about 3 to 6 degrees Celsius. The temperature could be more or less than this, depending on the amount of water you used and the amount of calcium chloride.

To keep track of your measurements, record them in the table shown below. You should record both the number of scoops of calcium chloride added and the *change in temperature* for each container. Give the thermometer time to adjust to each new temperature.

Container	Number of Scoops of Calcium Chloride	Temperature (°C)	Change in Temperature (°C)*
A			
B			
C			
D			

*Final temperature minus initial temperature

3. In container B add 2 level scoops of calcium chloride. Record the temperature of the water after the calcium chloride dissolved. Then calculate and record the change in temperature from the initial temperature of the water. Repeat the procedure for C and D, using 3 and 4 scoops of calcium chloride.

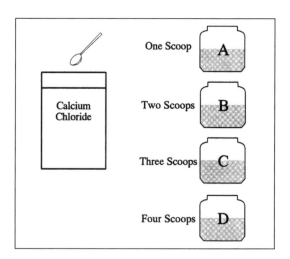

Your table of data should look similar to the one that follows except for the data itself. The initial temperature of the water we used was 20 °C. Your numbers will probably be different depending on the size of your scoop and other variables.

How does the amount of calcium chloride affect the temperature of H_2O?

Our Data

Container	Number of Scoops of Calcium Chloride	Temperature (°C)	Change in Temperature (°C)
A	1	25	5
B	2	33	13
C	3	41	21
D	4	43	23

Answer the following questions about what you did:

a. Did you use the same amount of water in each container? _____

b. Did you use the same amount of calcium chloride in each container? _____

c. Did the temperature change the same amount for each container? _____

d. What prediction would you make if you added five scoops of calcium chloride to the same amount of water? _____

Self-Check

If you followed the directions carefully, you used the same amount of water in all four containers. You added different amounts of calcium chloride to each container (1 scoop to container A, 2 to container B, 3 to container C, and 4 to container D). The temperature should have increased by different amounts in the containers. The increase was greater for those containers in which more calcium chloride was dissolved. Finally, you probably predicted that the temperature change would be even greater if five scoops of calcium chloride were added.

In this chapter you will learn to identify *variables* in a variety of contexts. Variables are those factors in an experiment that change or could potentially change. One type of variable is the one that the investigator purposely changes. This variable is called the *independent* variable. Another type of variable is the one that changes in response to changes in the independent variable. This variable depends on what is changed, and is therefore called the *dependent* variable.

Young students easily pick up the term *dependent* for the variable that responds to changes in the independent variable. However, they have more difficulty with the term *independent.* Independent is in the same linguistic form as *incomplete,* meaning *not complete.* Independent means *not dependent.* The term, therefore, tells us what it is not, rather than what it is.

We have found that young students will initially benefit from using other labels for these ideas. For example, some teachers call the independent variable the *manipulated* variable because the investigator purposely changes or manipulates that variable in an experiment. The dependent variable that responds could then be called the *responding* variable. Other teachers use the term *I change it* variable for the *independent variable.* Both begin with an '*i*' and the transition from the term *I change it* variable to *independent* variable may be easier. Although these teachers may initially use the *I change it* label for the independent variable, they use the term *dependent* variable right away because students already know what it means to depend on someone, and thus the label is more easily learned.

It is important to remember that as teachers we are teaching ideas, like variables that we purposely change and variables that change as a result. What we call those ideas is important, but that is an entirely different issue. In this book we will use the terms, *independent* and *dependent* variables because that is what scientists use. Young students ultimately will need to learn these terms as well because they are used on K–12 science standardized tests in many states. You will need to consult the science standards of your state to know the terms that students are expected to know.

Read this statement:

The height of bean plants depends on the amount of water they receive.

In this statement two variables are described—the height of plants and amount of water. A student interested in plant growth might purposely change the amount of water given to bean plants just to find out if their height was affected. Both *amount of water* and *height of bean plants* are factors that can vary or change. A variable is any factor in an experiment that can change.

Activity 7.2   **Learning to Identify Variables**

Read the following questions and record your responses. Answers can be found in the Self-Check that follows.

1. A **variable** is something that can vary or change. What are the variables in this statement?

   The time it takes to run a kilometer depends on the amount of daily exercise a person gets.

   _____   _____

2. What are the variables in this statement?

   The higher the temperature of water, the faster an egg will cook.

   _____   _____

3. What are the variables in this investigation?

   An investigation was conducted to see if keeping the lights on for different amounts of time each day affected the number of eggs chickens laid.

   _____   _____

4. Think back to the investigation you did at the beginning of this chapter. Then complete this statement:

   If the amount of calcium chloride added to the water increases, then the temperature of the water (increases, decreases).

   _____

5. What were the variables in the investigation on calcium chloride you conducted?

   _____   _____

6. If a variable is purposely changed, it is called an *independent* variable. It could also be called a *manipulated* variable, or even an *'I change it'* variable. Which of the two variables in the calcium chloride investigation was the *independent* variable?

   _____

7. What variable was manipulated in this investigation?

   The amount of pollution produced by cars was measured for cars using gasoline containing different amounts of additive.

   _____

8. What is the independent variable in this statement?

   Lemon trees receiving the most water produce the largest lemons.

   _____

9. Identify the independent variable in the following:

   The amount of algae growth in lakes seems to be directly related to the number of bags of phosphate fertilizer sold by the local merchants.

   _____

10. Identify the independent variable in the following:

    An investigation was performed to see if corn seeds would sprout at different times depending on the room temperature of the water in which they were placed.

    _____

Self-Check _____ Activity 7.2

1. amount of time to run a kilometer   amount of daily exercise

   It would not be correct to name just *time* or *exercise* as the variables. You must include how each variable will be measured or described. For example, height of plant, number of fruit produced, color of leaves, and diameter of the stem are all variables.

2. temperature of water   amount of time needed for an egg to cook

   Your answers do not have to be written exactly the same as these but they should be similar.

3. Number of hours (or amount) of light   number of eggs

   Again, your answers do not have to be exactly like these, but they should be similar. Just *light* and *eggs* however, would be incorrect descriptions of the variables.

4. increases

   As more scoops of calcium chloride were added, the more the water temperature increased.

5. number of scoops of calcium chloride   temperature change of water

    You might have said *amount of calcium chloride* or *temperature of water.* These answers would be correct also.

6. number of scoops of calcium chloride

    You purposely used a different number of scoops for each container so you manipulated this variable.

7. The *amount of additive in gasoline* is manipulated, so it is the independent variable.

8. The *amount of water* could be manipulated or changed to determine its effect on the size of the lemons produced.

9. The *number of bags of phosphate fertilizer sold* is the independent variable.

10. The *temperature of water* is the independent variable.

✔

In the previous questions, one variable was changed to see what would happen to another variable. A variable that is purposely changed is called an independent variable. The next series of questions will help you learn to distinguish between independent and dependent variables in a variety of investigative descriptions.

Activity 7.3   Distinguishing Between Independent and Dependent Variables

The more water you put on grass, the taller it will grow.

   Amount of water is the independent variable in the above statement. The other variable is the *height of grass.* The variable that may change as a result of changing the independent variable is called the dependent variable. Answers to the questions below can be found in the Self-Check that follows.

1. Identify the independent and dependent variables in this statement:

    More bushels of potatoes will be produced if the soil is fertilized more.

    independent variable: _____

    dependent variable: _____

2. Think back to the investigation you did at the beginning of this chapter. You manipulated the amount of calcium chloride. What was the responding variable?

    _____

3. Look at the sketch. It shows an investigation similar to the one you did. Notice that different amounts of water were used in each container. The amount of calcium chloride added to each container was kept the same. After the calcium chloride dissolved, the temperature change of the water in each container was determined.

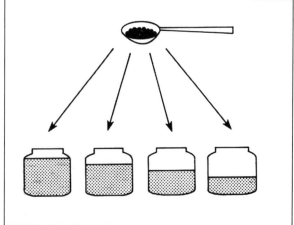

What are the independent (IV) and dependent (DV) variables in this investigation?

IV _____

DV _____

4. What are the independent (manipulated) and dependent (responding) variables in the following investigation?

Five groups of rats are fed identical diets except for the amount of Vitamin A that they receive. Each group gets a different amount. After three weeks on the diet, the rats are weighed to see if the amount of Vitamin A received has affected their mass.

IV _____

DV _____

5. An experiment was done with six groups of children to see if scores on their weekly spelling tests were affected by the number of minutes of spelling practice they had each day.

IV _____

DV _____

6. Will the number of nails picked up by an electromagnet be increased if more batteries are put in the circuit?

Suppose an investigation was conducted on the problem above. What would the variables be?

IV _____

DV _____

Self-Check ———————————————————————— Activity 7.3

1. *amount of fertilizer* independent (manipulated) variable
   number of bushels of potatoes dependent (responding) variable

   The amount of fertilizer could be *manipulated* to see if the number of bushels of potatoes *responded*.

2. *temperature change of water*

   In each container, you manipulated the amount of calcium chloride to see if the temperature of the water would respond.

3. (IV) *amount of water*

   (DV) *temperature change of water*

   This is similar to the investigation you did, but a different variable is being manipulated.

4. (IV) a*mount of Vitamin A*

   (DV) *mass of rats*

   If the *amount of Vitamin A* is manipulated or changed, then perhaps the *mass of rats* will respond. Of course, mass may not be affected if Vitamin A is not essential. The *mass of rats* is still the dependent variable whether or not it is actually affected by the independent variable.

5. (IV) *number of minutes of spelling practice,* (DV) *score on spelling test*

6. (IV) *number of batteries in circuit,* (DV) *number of nails picked up*

In this chapter, you learned to identify variables. You have also learned that a *variable* is something that can change or vary. A variable that is purposely changed is called an *independent* variable. The variable that may change as a result of changing the independent variable is called the *dependent* variable.

Now take the Self-Assessment for Chapter 7 and check your answers.

Continue on to Chapter 8 or return to Chapter 7 for more study based on the results of your Self-Assessment.

 # Self-Assessment

Identifying Variables

For each of the following statements or descriptions identify the independent variable (IV) and dependent variable (DV). The answers will follow.

1. Students in a science class conducted an investigation in which a flashlight was pointed at a screen. They wished to find out if the distance from the light to the screen had any effect on the diameter of the illuminated area.

    IV _____

    DV _____

2. The number of pigs in a litter is determined by the mass of the mother pig.

    IV _____

    DV _____

3. The State Agriculture Department has been counting the number of foxes in Brown County. Will the number of foxes have any effect on the rabbit population?

    IV _____

    DV _____

4. The score on the final test depends on the number of subordinate skills attained.

    IV _____

    DV _____

5. A study was done with white rats to see if the number of offspring born dead was affected by the number of minutes of exposure to X-rays by the mother rats.

    IV _____

    DV _____

Self-Assessment Answers

1. IV distance from light to screen
    DV diameter of illuminated area
2. IV mass of mother pig
    DV number of pigs in litter
3. IV number of foxes
    DV number of rabbits
4. IV number of subordinate skills attained
    DV score on final test
5. IV number of minutes of exposure to X-rays
    DV number of offspring born dead

High Stakes Testing

A sample multiple-choice item from State Standardized Exams.

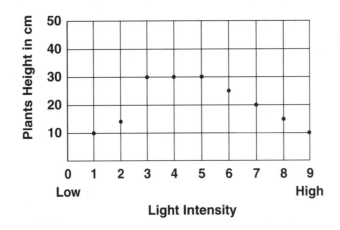

A student is investigating plant growth. The student can change the amount of light, heat, and water and the richness of the soil. During the investigations, the amounts of light, heat, and water remain the same. Only the richness of the soil changes. What is the [independent] variable in this experiment?

       A.   Light
       B.   Water
       C.   Heat
       D.   Soil ←

Illinois: ISAT Grade 4 Sample Test.

http://herrickses.org/searingtown/library/circleoflife/index.htm
ofcn.org/cyber.serv/academy/ace/sci/cecsci/cecsci132.html
www.studycoach.com/alg/classnotes/quadeqapp/idvarquadeqapp.html
www.kpbsd.kl2.ak.us/northstar/ns.staff/sm/sci.process.html

WEBSITES FOR ACTIVITIES

A Model for Assessing Student Learning

Assessment Type: Student Portfolio

Identifying variables is the first of several integrated science process skills students will learn to be able to eventually design and conduct their own experiments. Introducing student portfolios at this early point in instruction can evidence students' growth over time as they develop the necessary skills.

Two kinds of portfolios can be proposed to students—working folders and showcase folders. Working portfolios are holding bins for many pieces of work that is representative of learning still in progress. A showcase portfolio, on the other hand, evidences a student's best work. By periodically transferring pieces of a student's work from the working folder, the showcase portfolio gives the student, the parents, and the teacher a means to determine student performance levels based on his or her best work.

A showcase portfolio should include:

- a table of contents
- a student letter to the reviewer describing his or her portfolio
- a variety of entries that show what a student knows and is able to do.

Contents of a portfolio can include pretests, quizzes, tests, tables, graphs, laboratory reports, teacher notes, peer evaluations, and group designs of experiments. Because a portfolio is a collection of evidence, a science portfolio on the skills of experimenting should ultimately contain evidence of a student designed experiment. A written report of an experiment conducted at school or at home, or photographs of a science fair display of a science project can show mastery of the ability to design and analyze an experiment.

Portfolios encourage collaboration between teacher and students as they choose tasks and assignments that evidence their progress and achievement. Student self-evaluation is also fostered because students learn to monitor their own work and to reflect on their performance. The following are two examples of how self-evaluation can be encouraged.

**Science Research Checklist:
A Guide for Your Portfolio**

Provide evidence that you can do the following:

Planning
_____ I can choose my own topic.
_____ I can use several resources to find information on my topic.
_____ I can design or select an experiment based on my topic.

Conducting
_____ I can follow a procedure and use materials and equipment to conduct my experiment.
_____ I can collect data from my experiment.

Communicating
_____ I can display my data as tables and graphs.
_____ I can explain what I found.
_____ I can write a report describing my experiment.

Portfolio Evaluation

Name _____

Unit/Activity _____

Date _____

Why I chose this piece of work:
_____
_____
_____

What I learned:
_____
_____
_____

What I want to learn next:
_____
_____
_____

Chapter 8

Constructing a Table of Data

National and State Standards Connections --------------

- Use appropriate tools and techniques to gather, analyze, and interpret data. (NSES 5–8)
- Represent data using tables. (NCTM 3–5)
- Construct simple charts to display some of the information collected in the process of answering questions. (Delaware Content Standards, K–3)

Purpose

One of the skills needed to conduct an investigation is the organization of data in tables. When data are presented in well-organized tables, trends and patterns of change in data are often revealed.

Objectives

After studying the information in this chapter, you should be able to:

1. Construct a table of data when given a written description of the measurements made during an investigation.
2. Write data pairs from a table of data.
3. Match data pairs with points on a graph.

When an investigation is conducted, the measurements made are called data. Measurements of time, temperature, and volume are examples of data. Organizing data into tables helps to see patterns in the results.

This first set of exercises will acquaint you with how to set up a data table, assign appropriate labels to columns, and enter data.

Activity 8.1 Labeling Columns and Entering Data

Although there are no absolute rules for constructing tables of data, there are conventions, or commonly agreed upon patterns of organization, that facilitate communication between the writer and the reader. For example, when constructing a table of data, the independent variable is recorded in the left column and the dependent variable is recorded in the right column.

Column for the
independent variable   **Column for the**
                                  dependent variable

Answers for the following may be found in the Self-Check that follows.

1. Read the following description of an experiment and identify the independent and dependent variables. Record these variables in the table.

An investigation was conducted to see what happens to the height measured in centimeters of Wisconsin Fast Plants when 0, 1, and 2 of the plants' cotyledons were removed. (A cotyledon is the fleshy part of a seed, sometimes called a seed leaf.)

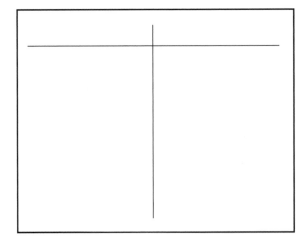

2. Which of the following tables have been properly ordered?

a. IV	DV		b. IV	DV		c. IV	DV		d. IV	DV		e. IV	DV
5	7		3	3		1	3		1	6		9	5
3	3		2	4		2	4		2	4		5	7
9	5		9	5		3	5		3	3		3	3
1	6		1	6		5	6		5	7		2	4
2	4		5	7		9	7		9	5		1	6

3. Try ordering the following data pairs in the table. The first number in each pair is for the independent variable and the second number is for the dependent variable.

(20, 17) (5, 18) (9, 12) (23, 26) (17, 3) (27, 32)

IV	DV

4. Try one more. Remember (IV, DV): (12, 12) (13, 10) (5, 10) (8, 12) (15, 8)

IV	DV

5. Here is a written description of an investigation. Notice that some of the data have already been recorded in a table. Read the paragraph and record the remaining data.

    The heights that balls bounced when dropped different distances were measured. A ball dropped 50 cm bounced 40 cm high. A 10 cm drop bounced 8 cm. A ball bounced 24 cm when dropped 30 cm. The bounce was 56 cm high for a 70 cm drop. A 100 cm drop bounced 80 cm.

    The Effect of Drop Height on Bounce Height

Height of Drop (cm)	Height of Bounce (cm)
10	8

6. Here is another practice problem. *Label* the columns and *record* the data in the table provided.

    The distance covered by a runner during each second of a race was measured. During the 15th second of the race the runner covered two meters. Three meters were covered during the 12th second. Four meters were covered during the 9th second. During the 6th second three meters were covered. During the 3rd second, two meters were traveled.

    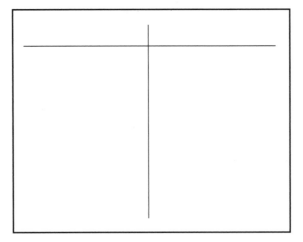

Self-Check _____ Activity 8.1

1. **How Does Removing Cotyledons Affect the Height of Plants?**

Number of Cotyledons Removed	Height of Plants (cm)

Notice that whenever units are used, they are included in the column heading as well.

When recording data in a table, the levels of the independent variables are ordered. Although data are sometimes ordered from largest to smallest, the usual procedure is to order data from smallest to largest. This organization establishes a pattern of change in the independent variables. If there is a corresponding pattern of change in the dependent variable, it will be easier to recognize than if the levels of the independent variable were placed randomly in the table.

2. Only d. and e. are correct. In Table a. the data are in random order and therefore incorrect. In b. the data for the dependent variable have been ordered. This is not correct, unless it occurs naturally as a result of ordering the data of the independent variable. In c. the data pairs were separated, which is incorrect, and then ordered. Table e. is correct, but the usual procedure is to order the data from the smallest to the largest as in Table d.

3. Either is correct, but the data for the independent variables are usually ordered from the smallest to the largest.

IV	DV
5	18
9	12
17	3
20	17
23	26
27	32

IV	DV
27	32
23	26
20	17
17	3
9	12
5	18

4. Again, either is correct, but the first is preferred, where the data are ordered from the smallest to largest.

IV	DV
5	10
8	12
12	12
13	10
15	8

IV	DV
15	8
13	10
12	12
8	12
5	10

5. The Effect of Drop Height on Bounce Height

The Effect of Drop Height on Bounce Height

Height of Drop (cm)	Height of Bounce (cm)
10	8
30	24
50	40
70	56
100	80

When constructing tables of data, the independent variable, Height of Drop, heads a column on the left, with all the levels listed below, usually from smallest to largest. The dependent variable is recorded in the column to the right. Units of measurements are placed in the column heading and not next to the data.

6.

Time During Race (sec)	Distance Covered (m)
3	2
6	3
9	4
12	3
15	2

Activity 8.2 Modifying a Table of Data to Record Repeated Trials

Most experiments should be repeated by testing each level of the independent variable several times. Repeated trials *increase confidence* in results by reducing the effects of chance errors that may occur in a single trial.

When repeated trials are conducted, the column for the dependent variable is divided into smaller columns so data can be recorded for each repeated trial. Information, such as the average result (mean, median, or mode) or the range (how spread out the data are), is recorded in one or more columns to the right of the column for the responding variable. Information that is computed from data is called a *derived quantity*.

	Trials			
	1	2	3	

Column for the Independent Variable   **Columns for the Dependent Variable**   **Column for a Derived Quantity**

1. Read the following description of an investigation and construct a table of data:

   - Draw columns for the independent variable, dependent variable, and a derived quantity.
   - Subdivide the column for the dependent variable to reflect the number of repeated trials conducted.
   - Write labels and appropriate units for each column.

An experiment was conducted to see how the number of minutes of heating time affected the temperature of water. A pan containing 1 liter of water was heated for 5 minutes. At the end of each minute the temperature in degrees Celsius was recorded. The experiment was conducted a total of 4 times. Mean *(arithmetic average)* temperatures were calculated for each level of the independent variable.

Self-Check _____ Activity 8.2 ✔

1.   **How Does Heating Time Affect the Temperature of Water?**

Length of Heating Time (min)	Temperature (°C)				Mean Temperature of Water (°C)
	Trials				
	1	2	3	4	

You have just learned to set up a data table and assign labels to the columns for data. A data table organizes data and helps us to recognize patterns and trends in the data.

In this next set of exercises you will learn to transfer data from a table to a graph. A graph is a visual representation of the data in a table. The word graph comes from a Greek word meaning picture.

Activity 8.3   Transferring Data from a Table to a Graph

Transferring data from a data table to a graph is an easy 2-step process in which you:

1.   observe the data in the data table and write number pairs
2.   use the number pairs to locate data points on a graph.

Because graphs communicate data in pictorial form, they can show trends and patterns in data more effectively than a data table alone. Each point on a graph represents a pair of data. When writing data pairs, the value for the horizontal or X axis on a graph is written first, followed by the value for a graph's vertical or Y axis. The two numbers are separated by a comma and are placed in parentheses, for example (10, 18). By convention, the independent variable is graphed on the horizontal (X) axis and the dependent variable on the vertical (Y) axis.

If the convention for the placement of the IV and DV in data tables is followed, each data pair for a graph (IV, DV) can easily be determined from a table.

1. Answers for the following may be found in the Self-Check that follows. Write the data pairs for constructing a graph from the following table of data.

The Effect of Amount of Rain on the Mass of Fruit

Amount of Rain (cm)	Mass of Fruit (kg)
45	125
55	140
60	150
66	200
70	280
75	310

(,   )

(,   )

(,   )

(,   )

(,   )

(,   )

A data point on a graph is described using the values given on the horizontal and vertical axes. Imagine a **vertical** line drawn to the point labeled "F' on the graph. Such a line would intersect the X axis at the point with a value of 25 (halfway between the intervals of 20 and 30). Imagine a second line drawn **horizontally** to point "F." This line would intersect the Y axis at a point with a value of 10.

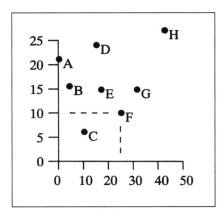

2. Match the letters of the remaining data points with the following number pairs:

(0, 21) _____   (18, 15) _____

(5, 16) _____   (25, 10) _____

(10, 6) _____   (31, 15) _____

Self-Check _____ Activity 8.3 ✔

1. (45, 125)
    (55, 140)
    (60, 150)
    (66, 200)
    (70, 280)
    (75, 310)

2. (0, 21) _____ A _____   (18, 15) _____ E
    (5, 16) _____ B _____   (25, 10) _____ F
    (10, 6) _____ C _____   (31, 15) _____ G

Being able to write data pairs from a table of data is the first step in being able to construct a graph. In the next chapter you will learn to make a graph from a data table.

Now take the Self-assessment for Chapter 8.

Self-Assessment

Constructing a Table of Data

1. Construct an ordered table of data for the following. The lengths of shadows made by sticks of different length were measured. A stick 50 cm long made a shadow 40 cm long. The shadow was 5 cm long for a stick 10 cm. A 30 cm stick made a shadow 22 cm long and a stick 40 cm long made a shadow of 29 cm. The shadow was 12 cm for a stick 20 cm long.

2. Construct a table to record data for this investigation. Include smaller columns for repeated trials as appropriate.

   A study was conducted to see if amount of salt affected how fast the salt dissolved in water. Using the same amount and temperature of water, four different amounts of salt (10, 20, 30, and 40 mL) were stirred until no crystals were visible. Each amount of salt was tested three times.

3. Match the letter with the data pair that describes the location of each point.

How Does Length of Stick Affect Length of Shadow?

Length of Stick (cm)	Length of Shadow (cm)

Data Pairs	Letter
(3, 26)	_____
(5, 42)	_____
(15, 10)	_____
(17, 45)	_____
(20, 25)	_____
(30, 50)	_____

Self-Assessment Answers

1.

How Does Length of Stick Affect Length of Shadow?

Length of Stick (cm)	Length of Shadow (cm)
10	5
20	12
30	22
40	29
50	40

2. Amount of salt (mL)

The Effect of Amount of Salt on the Time to Dissolve

Amount of Salt (mL)	Time to Dissolve (min)			Mean Time to Dissolve (min)
	Trials			
	1	2	3	
10				
20				
30				
40				

3.

Number Pairs	Letter
(3, 26)	B
(5, 42)	A
(15, 10)	C
(17, 45)	D
(20, 25)	E
(30, 50)	F

High Stakes Testing

A sample multiple-choice item from State Standardized Exams.

Trial	Temperature (°C)
1	41
2	40
3	31
4	42

Students conducted an experiment in which they rubbed their palms together to warm their hands, then measured the temperature of their hands. The experiment was conducted 4 times. According to the data in the table, which of these trials is most unusual?

    F.   1
    G. 2
    H. 3 ←
    I.   4

Virginia: SOL Grade 5 Spring 2001 Release Item.

askeric.org/Virtual/Lessons/Mathematics/Probability/PRB0005.html
pittsford.monroe.edu/jefferson/calfieri/graphs/Tables.html
java.sun.com/docs/books/tutorial/uiswing/components/table.html
cstl.syr.edu/fipse/TabBar/Buildtbl/buildtbl.htm

WEBSITES FOR ACTIVITIES

A Model for Assessing Student Learning

Assessment Type: Student/Peer/Family Checklist

Name: _____ Date: _____

Data Table Title: _____

Criteria	Student		Peer		Family	
Tabulating Data Skills	Yes	No	Yes	No	Yes	No
Does the title tell about the independent variable (IV) and the dependent variable (DV)?						
Is the left-hand column for the IV?						
Are the label and units given for the IV?						
Are the levels of the IV ordered?						
Is the right-hand column for the DV?						
Are the label and units given for the DV?						
Is the DV column subdivided for repeated trials?						
Are the DV data correctly recorded?						
Are there additional columns for derived quantities such as the mean and the range?						
Are the label and units given for the derived quantities?						
Are the derived quantities correctly calculated?						

Constructing a Graph

Purpose

A picture is worth a thousand words. Graphs communicate in pictorial form the data collected from experiments and other sources. Usually, graphs communicate patterns and information better than data tables. However, graphs are more difficult to construct and involve several sub-skills, including knowledge of the major parts of a graph, translating data pairs from a data table to points on a graph, constructing an appropriate scale for each axis, plotting the data on a graph, and finally, summarizing the trends through a line-of-best-fit and descriptive sentences. By using a series of steps, you can easily construct the graph you need to display your data. There are many kinds of graphs, including bar graphs, histograms, scatter plots, and line graphs. Chapters 9 and 10 will help you learn to construct *line graphs* as well as provide you with a model of how to teach students to construct a graph.

Objective

After studying this chapter you should be able to construct a graph when provided with a brief description of an investigation and a table of data.

An example of what you will be able to do when you finish this chapter is shown below. You will be given the type of information found on the left-hand side of the page and will be expected to produce the type of material found on the right-hand side.

You will be given:

INVESTIGATION: Beans were soaked in water for different lengths of time and their gain in mass was recorded.

Amount of Soaking Time (min)	Mean Gain in Mass (g)
5	10
10	20
15	40
20	45
25	50
30	55

You will produce:

The Effects of Soaking Time on Mass of Beans

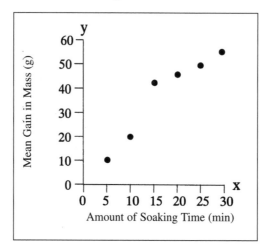

The part of the graph above, labeled *Mean Gain in Mass,* is called the vertical or **Y** axis and that part labeled *Amount of Soaking Time* is called the horizontal or **X** axis. Together they make up the axes of the graph.

In order to construct a graph from a table of data you must learn three skills:

1. Label the X axis with the independent (manipulated) variable and the Y axis with the dependent (responding) variable.
2. Determine an interval scale for each axis that is appropriate for the data to be graphed.
3. Plot the data pairs as data points on a graph.

Activity 9.1 Skill 1: Labeling the X and Y Axes

Amount of Fertilizer (kg)	Height of Plants (cm)
2	24
4	50
6	74
8	38

Suppose you wanted to construct a graph of the data given above. What would you have to do? After drawing the horizontal and vertical axes, write labels for the variables along these axes. When deciding which variable to assign to an axis, follow this rule:

The independent (manipulated) variable is written along the horizontal or X axis, while the dependent (responding) variable is written along the vertical or Y axis.

Although there are exceptions, this general rule helps the reader quickly interpret a graph by knowing which variable has been purposely changed and which responds as a result. Placing the independent variable on the horizontal axis and the dependent variable on the vertical axis results in up and down variation, which is easier to interpret as well.

In the case shown above, the amount of fertilizer was purposely manipulated and the height of the plants was then measured. So, *Amount of Fertilizer (kg)* should be written along the horizontal axis and *Height of Plants (cm)* should be written beside the vertical axis. The correct form is shown on the graph below:

The Effect of Amount of Fertilizer on Height of Plants

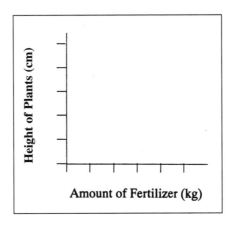

Descriptions of several investigations follow. Next to each description is a graph with the variables assigned to the axes. Your task is to determine whether the labels are correct. The answers are found in the Self-Check that follows.

1. *INVESTIGATION:* A ball is dropped from several distances above the floor and the height it bounces is then measured.

❏ Labels are correctly placed

❏ Labels are reversed

How Does the Distance from Which a Ball Is Dropped Affect How High It Bounces?

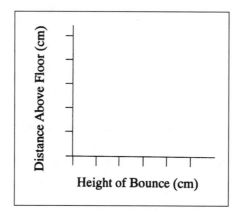

2. *INVESTIGATION:* A candle was burned under glass jars of different volumes to see if the length of time the candle burns is affected by the volume of the jar.

❏ Labels are correctly placed

❏ Labels are reversed

How Does the Volume of a Jar Affect the Burning Time of a Candle?

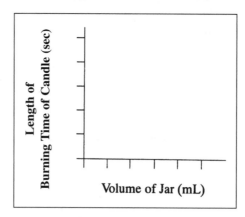

3. *INVESTIGATION:* An investigation is done to see if the diameter of rubber tubing affects the length of time it takes to siphon water out of a container.

❑ Labels are correctly placed

❑ Labels are reversed

How Does the Diameter of a Siphon Affect Siphoning Time?

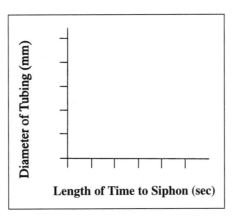

Several descriptions of investigations follow. For each graph write the independent (manipulated) variable along the horizontal or X axis. Write the dependent variable along the vertical or Y axis. Be sure to include the appropriate measurement units in parentheses for each variable.

4. *INVESTIGATION:* A fisherman used fishing lines of several different gauges and recorded the number of fish caught on each.

Gauge of Line (test pounds)	Number of Fish Caught
6	1
8	5
10	12
12	20
15	37
20	22

How Does the Gauge of Fishing Line Affect the Number of Fish Caught?

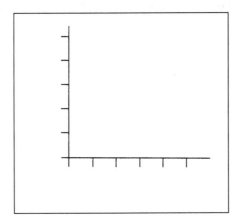

5. *INVESTIGATION:* A study was conducted to see if the number of surfers on the beach was affected by the height of the waves.

Height of Waves (m)	Number of Surfers
1	13
2	23
3	56
4	31

How Does the Height of the Waves Affect the Number of Surfers on the Beach?

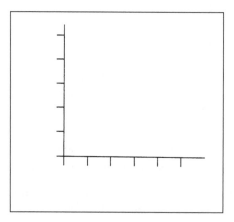

6. *INVESTIGATION:* Rocks from several different depths in a mine were collected. The density of each rock was recorded.

Depth of Collection (m)	Density of Rocks (g/cm^3)
0	2.2
30	2.0
120	2.7
600	3.5
3000	4.0

How Does the Depth of Collection Affect the Density of Rocks?

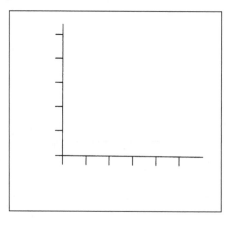

Self-Check ———————————————————— Activity 9.1

1. Labels are reversed.
   Because the ball was deliberately dropped from different heights, that variable is manipulated and should be used to label the horizontal axis.
2. Labels are correctly placed.
3. Labels are reversed.
4. **How Does the Gauge of Fishing Line Affect the Number of Fish Caught?**

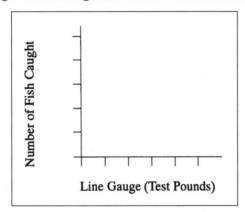

5. **How Does the Height of the Waves Affect the Number of Surfers on the Beach?**

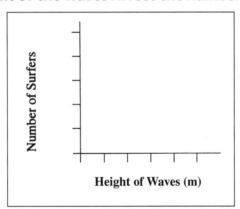

6. **How Does Depth of Collection Affect the Density of Rocks?**

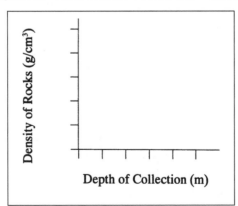

Remember, by convention, the independent (manipulated) variable is written in the left column of a data table. Read the description of an investigation to decide which variable has been manipulated.

Activity 9.2   Skill 2: Determining Interval Scales for Each Axis

Refer to this data table for the following example:

7	
12	
22	
37	
46	
55	

Finding the right scale for numbering the axes of a graph can be challenging. Trial and error is one approach that may help find the right scale for each axis. However, most students are not very good with this approach; they need a little structure to guide them. An easy way to find a good scale to fit the data consists of the following steps:

Steps	**Example**
1. Find the range of the data to be graphed by subtracting the smallest value from the largest value. 2. Divide this difference by the number of intervals you want. If you want about 5 intervals, divide by 5. This usually results in a scale with 5 to 7 intervals. Too many intervals crowd a graph, while too few make it difficult to plot data points. 3. Using 9.6 to make intervals would be awkward, so to make the job easier round 9.6 to an easy counting number like 10. Good counting numbers are usually multiples of 5, like 10 or 20, or smaller numbers like 2 and 4. 4. Use this rounded number (10 in this example) to mark off the intervals along the axis. Begin with a multiple of 10 that is less than the smallest value to be plotted (7 in this example) and continue until you have reached or exceeded the largest value to be plotted. Numbering for both axes always begins at the intersection of the axes, called the origin. These steps result in a scale that uses the entire graph area.	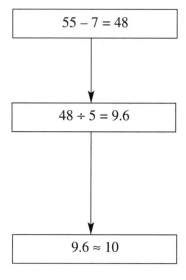

For questions 1–6, refer to the following table. For each question, determine if the interval scale is correct. The answers can be found in the Self-Check that follows.

Cost of Watch ($)	Error per Month (min)
249	2
225	4
220	5
200	6
124	9
110	10

1.
❏ A. Scale is correct
❏ B. Sizes of intervals are not equal
❏ C. Too many intervals
❏ D. Too few intervals
❏ E. Starts with too small an interval

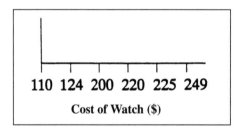

2.
❏ A. Scale is correct
❏ B. Sizes of intervals are not equal
❏ C. Too many intervals
❏ D. Too few intervals
❏ E. Starts with too small an interval

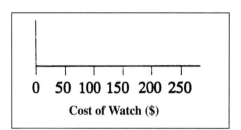

3.
❏ A. Scale is correct
❏ B. Sizes of intervals are not equal
❏ C. Too many intervals
❏ D. Too few intervals
❏ E. Starts with too small an interval

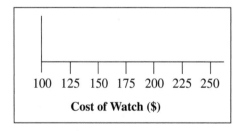

4.
❏ A. Scale is correct
❏ B. Sizes of intervals are not equal
❏ C. Too many intervals
❏ D. Too few intervals
❏ E. Starts with too small an interval

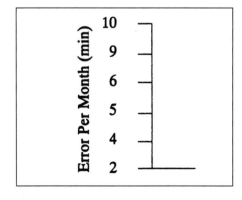

5.
- ❏   A.   Scale is correct
- ❏   B.   Sizes of intervals are not equal
- ❏   C.   Too many intervals
- ❏   D.   Too few intervals
- ❏   E.   Starts with too small an interval

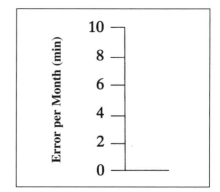

6.
- ❏   A.   Scale is correct
- ❏   B.   Sizes of intervals are not equal
- ❏   C.   Too many intervals
- ❏   D.   Too few intervals
- ❏   E.   Starts with too small an interval

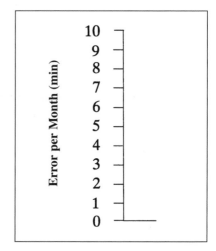

Self-Check _____ Activity 9.2 ✔

1. **E,** starts with too small an interval. By starting with 0 only about half of the available space on the graph is used. The first number on the axis should be the smallest number to be graphed or some number only slightly smaller. It is not necessary to start with 0. For instance, in this problem the first number should have been 100 or some number smaller than 110.

2. **B,** sizes of the intervals are not equal. The difference between 110 and 124 is not the same as the difference between 124 and 200. *The size of the intervals must be equal.*

3. **A,** scale is correct. It begins with an interval just below the smallest value and continues with equal intervals of 25 to accommodate all the data.

4. **B,** sizes of intervals are not equal.

5. **A,** scale is correct.

6. **C,** too many intervals. Excessive intervals clutter the scale; 5 or 6 intervals would be better.

You have just had some practice in distinguishing between appropriate and inappropriate numerical scales. In questions 7-10, you will use your skills to determine scales on your own.

7. Mark the horizontal axis of the following graph outline into five or so equal line segments. Use a series of small marks along the axis. Mark the vertical axis into about five equal line segments as well. Label each mark on the horizontal axis in intervals of 5. Begin the interval scale with 35.

Examine the table of data below. Determine the range of the data for *Height of Plant (cm)* by subtracting the smallest value from the largest value. The difference between 36 and 57 is 21. Divide by the number of desired intervals (we chose 5) to determine the size of the intervals. When 21 is divided by 5, the result is 4.2. Round 4.2 to a convenient counting number such as 5. The smallest number you must graph is 36. Begin labeling the axis with a multiple of 5 that is less than 36 (35). Some people think that all graphs begin with the intersection of the X and Y axes labeled as (0, 0). Graphs that begin at (0, 0) work fine when the data sets start at or near zero. But in the real world and in many experiments, data sets often begin with numbers far from zero, resulting in a large gap between 0 and the first piece of data to be graphed. If you are uneasy about not beginning at zero, you may wish to use the symbol (-//- or $\not=$) as part of the axis to indicate that part of the graph is not shown. Many teachers include this option as part of their instruction on graphing data.

How Does the Height of a
Plant Affect the Number of Leaves?

Height of Plant (cm)	Number of Leaves
36	68
42	73
47	90
50	180
53	116
57	216

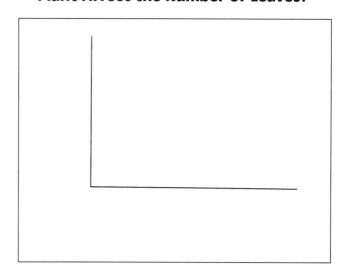

Now let's repeat the procedure on the data for the dependent (responding) variable, *Number of Leaves*. Find the range of data to be graphed by calculating the difference between the largest and smallest values for this variable, 216 – 68 = 148. Divide this range by 5 to determine the size of the interval, 148 ÷ 5 = 29.6. (Remember, we divide the range by 5 because it results in a reasonable number of intervals for most data sets.) At this point you have the choice of rounding 29.6 up to 30 or down to 25. Both are good counting numbers and either would be acceptable. Begin labeling the scale with a multiple of 30 (or 25) that is less than the smallest piece of data to be graphed (68).

8. Label the vertical axis for the blank graph above using your choice of intervals of 25 or 30.

9. Here is a second problem. Label the axes with appropriate numerical scales for graphing the data in the table below. Remember to mark off each axis with about 5 marks and then devise an appropriate numerical scale for each.

Length of Time from Start to Hatching (hours)	Number of Flies Hatched
9	25
12	153
14	269
15	617
18	1245

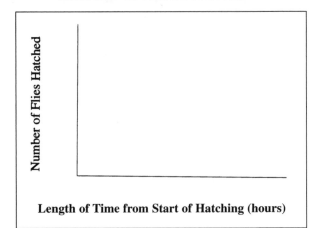

The Effect of Hatching Time on Number of Flies Hatched

10. Try one more problem on determining an interval scale. Label the axes with appropriate scales for graphing these data.

Width of Mower (cm)	Length of Time to Mow Field (min)
240	37
210	43
175	110
160	125
143	180

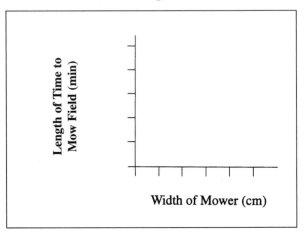

The Effect of Mower Width on Mowing Time

Self-Check

7.

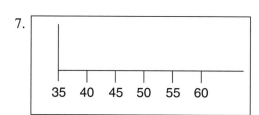 or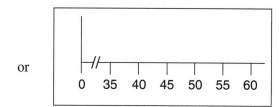

8.

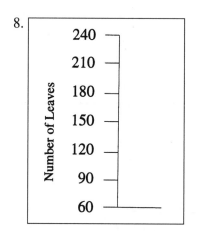 or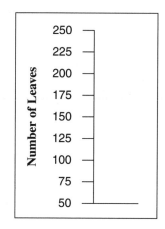

9.

The Effect of Length of Hatching Time on Number of Flies Hatched

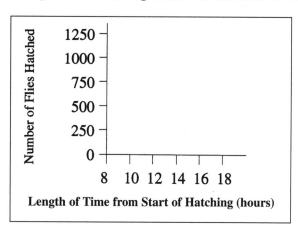

10.

The Effect of Mower Width on Mowing Time

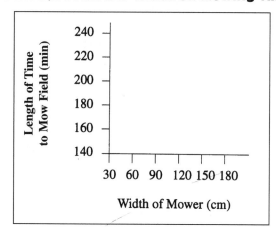

Activity 9.3   Skill 3: Plotting Data Pairs as Points on a Graph

You should now be ready for the third skill you need to construct a graph. After you have labeled the variables along the X and Y axes and have determined appropriate scales for each axis, you are ready to plot points for data pairs in a table of data. The answers can be found in the Self-Check that follows.

1. Refer to the following table of data. The first data pair is (8,6). On the graph locate 8 on the horizontal axis and 6 on the vertical axis. Imagine a vertical line drawn straight up from the 8 and a horizontal line drawn straight across from the 6. Where these two a imaginary lines intersect is a point representing that data pair.

    Look at the second pair of data in the table (10, 15). Sight imaginary lines straight up from 10 and straight across from 15. The point at which these imaginary lines intersect is a point representing the data pair, (10, 15).

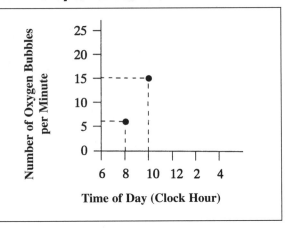

The Effect of Time of Day on Number of Oxygen Bubbles Released by a Plant in Water

Time of Day (Clock Hour)	Number of Oxygen Bubbles per Minute
8	6
10	15
12	27
2	19
4	5

2. Here is another table of data. The positions of the first two data pairs from this table are already plotted on the graph. You plot the remaining four data pairs.

Temperature of Freezer (°C)	Length of Time to Freeze (min)
−27	14
−20	20
−13	30
−8	43
−3	55
0	65

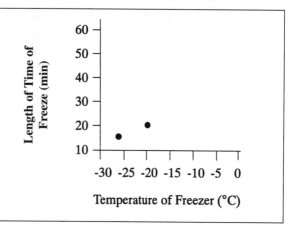

How Does the Temperature of the Freezer Affect Freezing Time?

3. Plot the data pairs on the graph using this table of data.

Distance from Bulb (cm)	Temperature of Air (°C)
5	55
10	40
15	31
20	28
25	25

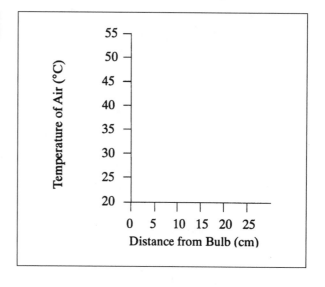

How Does Distance from a Light Bulb Affect Air Temperature?

4. Plot the data points for this table on the graph.

How Does the Date Affect the Number of Library Books Checked Out?

Date in November	Number of Books Checked Out
8	675
13	353
15	430
20	270

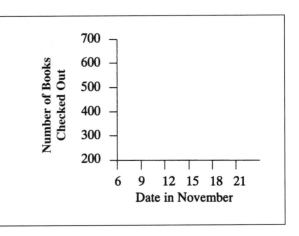

Self-Check ————————————————————————— Activity 9.3

1. The Effect of Time of Day on Number of Oxygen Bubbles Released by a Plant in Water

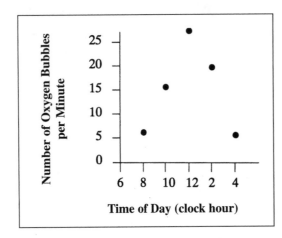

2. How Does the Temperature of the Freezer Affect Freezing Time?

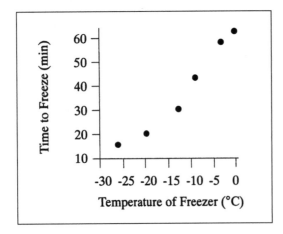

3. How Does Distance from a Light Bulb Affect Air Temperature?

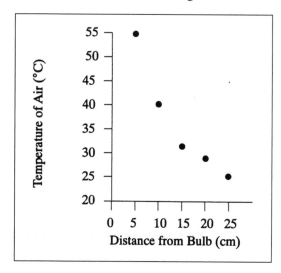

4. How Does the Date Affect the Number of Library Books Checked Out?

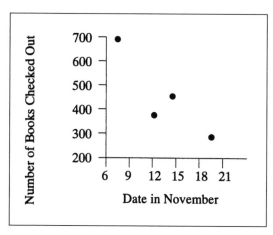

You have just practiced the third skill you will need in order to construct a graph. All you have to do now is put together the skills you have learned and you will be able to construct a graph.

Remember, for a graph to be properly constructed you must be able to do three things:

1. Write labels for the variables along the correct axis.
2. Determine an appropriate interval scale for each axis.
3. Plot each data pair as a data point on the graph.

Activity 9.4 Critiquing Graphs

In the last part of this chapter you will examine several graphs to see if they have been constructed correctly. Then when given a description of an investigation and a table of data, you will be expected to construct your own graphs.

An investigation and a table of data follow. Two graphs of the data are shown. For each graph determine if it is constructed correctly. If you think the graph is not correct, check the reason it is incorrect. The answers for the following can be found in the Self-Check that follows.

INVESTIGATION: The temperature of the air was measured at several times during the day.

The Effect of Time of Day on Temperature of Air

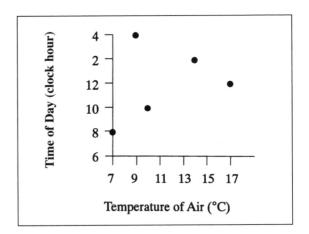

Time of Day (clock hour)	Temperature of Air (°C)
8	7
10	10
12	17
2	14
4	9

1.
- ❏ A. Graph is correct
- ❏ B. Variables are on wrong axes
- ❏ C. Incorrect scale
- ❏ D. Data points are in wrong location

The Effect of Time of Day on Temperature of Air

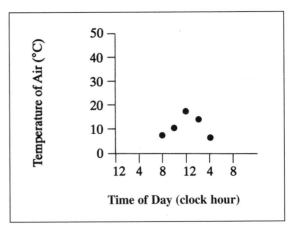

2.
- ❏ A. Graph is correct
- ❏ B. Variables are on wrong axes
- ❏ C. Incorrect scale
- ❏ D. Data points are in wrong location

3. Now try putting all three skills together. Construct a graph of data in this table.

   INVESTIGATION: The number of kilometers per liter of gasoline was measured for cars traveling at different speeds.

How Does Car Speed Affect Gasoline Consumption?

Speed of Car (km/h)	Kilometers per Liter
20	6.0
25	5.5
30	4.0
35	3.7
40	3.5

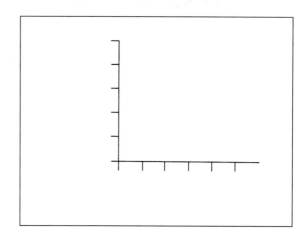

4. Try another problem. Construct a graph for the data from this investigation.

   INVESTIGATION: The mean mass of ten pumpkins growing in a patch was determined at different times after planting.

The Effect of Planting Time on Pumpkin Mass

Length of Time after Planting (weeks)	Mean Mass of Pumpkins (kg)
2	0
7	0
9	1
12	9
16	15
18	22

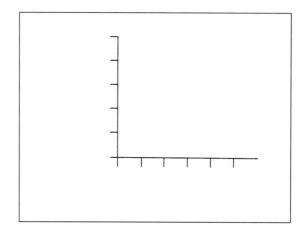

5. *INVESTIGATION:* A box was dropped from an airplane and the distance it had fallen was measured after various lengths of time.

Length of Time (sec)	Distance Fallen (m)
1	5
2	20
3	45
4	80
5	125

The Effect of Time on Distance Fallen

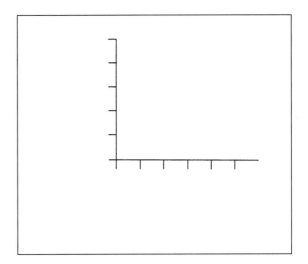

Self-Check ——————————————————— Activity 9.4

1. **B,** variables are on wrong axes. The labels are reversed. The *time of day* is the independent (manipulated) variable so it should be on the horizontal axis.
2. **C,** incorrect scale. The numerical scales on both axes are too large. Notice that the data points are all in a small clump. With appropriate scales the points should be distributed throughout the graph.

3.

How Does Car Speed Affect Gasoline Consumption?

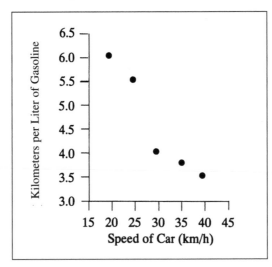

4. The Effect of Planting Time on Pumpkin Mass

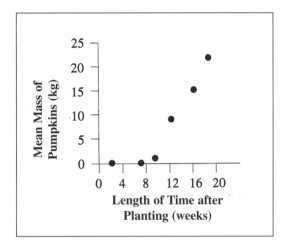

Note: Compare the origin (the intersection of the X and Y axes) of this Self-Check (0, 0) with the previous Self-Check for question # 3 (3.0, 15). You should be able to explain why one graph begins with zero on each axis, while the other graph does not. If not, re-read page 169 for an explanation.

5. The Effect of Time on Distance Fallen

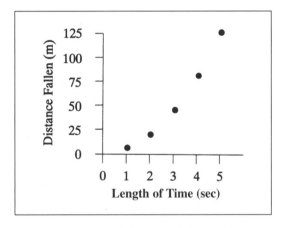

You have just learned to construct a graph from the data in a table. This is a valuable skill you will use later in this book and, hopefully, with your students. In Chapter 10 you will learn to draw a line-of-best-fit and to write a summary description of a graph.

Now take the Self-Assessment for Chapter 9.

Self-Assessment

Constructing a Graph

1. a. Write labels for the variables from this investigation along the appropriate axes of the graph.

    INVESTIGATION: The temperature of the water was varied in several containers to see if the length of time to evaporate the water was affected.

    b. What is the rule used to label the axes of a graph?

How Does Water Temperature Affect Evaporation?

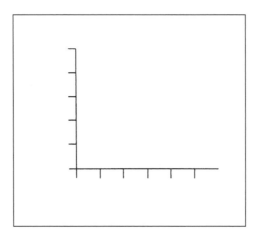

2. Label the axes with appropriate interval scales for graphing these data. You do not have to plot the points.

    INVESTIGATION: The temperature of the air was measured on different days to see if it affected the number of swimmers on the beach.

Temperature of Air (°C)	Number of Swimmers
12	30
19	80
20	225
26	450
31	475

How Does the Air Temperature Affect the Number of Swimmers?

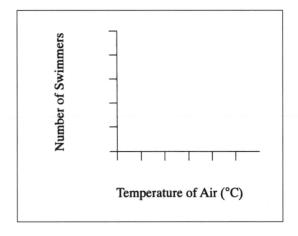

3. Plot the data pairs from this investigation on the graph.

    INVESTIGATION: The number of letters that could be correctly identified on an eye chart at different distances was investigated.

Distance of Eye from Chart (m)	Number of Letters Identified
1	18
2	26
3	34
4	30
5	22

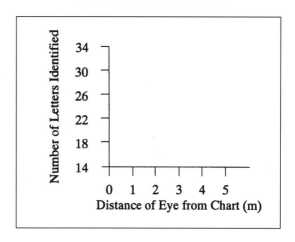

The Effect of Eye Distance on Letters Identified

4. Construct a graph of the data in this table.

    INVESTIGATION: Ice cubes of different sizes are melted in a pan of water.

Mass of Ice Cubes (g)	Length of Time to Melt (min)
35	2
45	3
52	5
61	9
70	11

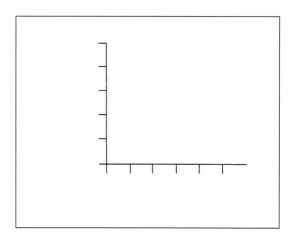

How Does the Mass of Ice Cubes Affect Melting Time?

Self-Assessment Answers

1. a.

How Does Water Temperature Affect Evaporation?

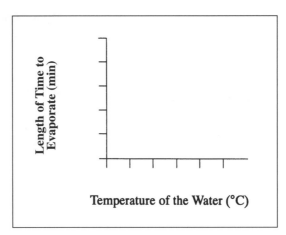

b. Rule: The independent (manipulated) variable is written along the horizontal axis.

2.   How Does the Air Temperature Affect the Number of Swimmers?

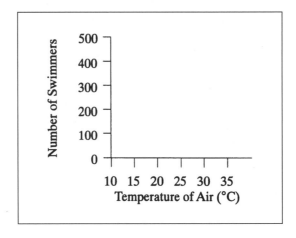

3. **The Effect of Eye Distance on Number of Letters Identified**

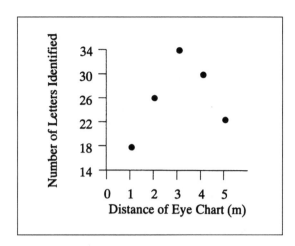

Note: When you determined an appropriate scale for the vertical axis (34 − 18 = 16; 16 ÷ 5 = 3.2), you could have rounded down to 3, or up to 4 or even 5. Intervals of 3, 4, or 5 could be used with this data.

4. **How Does the Mass of Ice Cubes Affect Melting Time?**

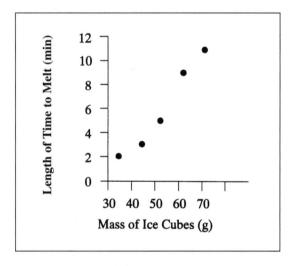

Note: To be correct:

a. The variables must be written along the axes as shown.

b. The interval scales should be about the same as these. If you started the interval scale with 0, be sure you used the symbol (-//-) to indicate that part of the graph is missing. It would be inappropriate to label the horizontal axis from 0 to 70 in intervals of 10 because much of the graph would be blank and the data would be shifted to the right.

High Stakes Testing

A sample multiple-choice item from State Standardized Exams.

The graph below shows the distance and time traveled by four cars.

Distance and Time Traveled

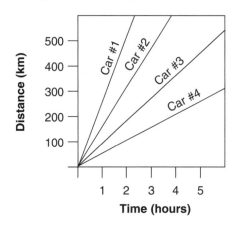

Which car traveled the slowest?

1. *Car #1*
2. *Car #2*
3. *Car #3*
4. *Car #4*

New York: Intermediate Level Science Sampler Spring 2000.

www.mste.uiuc.edu/courses/ci330ms/youtsey/intro.html
www.eduplace.com/ss/act/graph.html
www.exploratorium.edu/learning_studio/ozone/graphing1.html
http://viz.globe.gov/viz-bin/home.cgi

WEBSITES FOR ACTIVITIES

A Model for Assessing Student Learning

Assessment Type: Teacher Rating Sheet

Name: _____ **Date:** _____

Graph Title: _____

Criteria	Teacher Rating	
Graphing Skills	**Possible Points**	**Earned**
Does the title communicate the independent variable and dependent variable?	5	
Is the independent variable on the X axis?	10	
Are the label/units given for the independent variable?	10	
Is the scale on the X axis appropriate to represent the values of the independent variable?	10	
Is the dependent variable on the Y axis?	10	
Are the label/units given for the dependent variable?	10	
Is the scale on the Y axis appropriate to represent the values of the dependent variable?	10	
Are the data correctly plotted?	15	
Is the line-of-best-fit appropriate?[1]	10	
Is the graph done neatly?	10	

[1]Discussed in Chapter 10

Describing Relationships between Variables

National and State Standards Connections

- Think critically and logically to make relationships between evidence and explanations. . . . form logical arguments about the cause-and-effect relationships in an experiment. (NSES 5–8)
- Investigate how a change in one variable relates to a change in a second variable. (NCTM 3–5)
- Recognize simple patterns in data and use data to create a reasonable explanation for the results of an investigation or experiment. (Massachusetts Curriculum Framework, 3–5)

Purpose

In Chapter 9 you learned three major skills associated with constructing a graph. In this chapter you will learn two additional skills that are needed to interpret a graph: drawing lines-of-best-fit and writing sentences that summarize trends.

Objectives

After studying the information in this chapter you should be able to:

1. Draw a best-fit line when given a graph.
2. Describe in writing the relationship between variables on a graph.

You might think of a graph as a coded message; it means a great deal to the person who understands the code but not much to anyone else. Interpreting a graph begins with looking for patterns or trends in the data points. After plotting all the data points, look for a pattern in the points. Is there a general upward trend of the points? Or is the trend downward? Do the points go up and then level off? Or do they gradually increase, reach a peak, and then gradually decrease?

After you have identified a pattern or trend in the data points, try drawing a line that best shows that trend. Do not make a zigzag line that just connects the points. You are trying to draw a line that shows the general trend of the data. Draw the line so that about half the points are on one side of the line and half are on the other. Some points may actually be on the line. Examples of lines-of-best-fit are shown below.

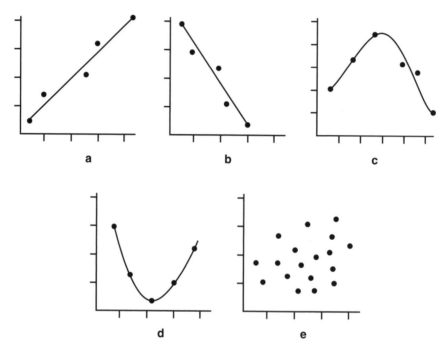

Figure 10.1 Examples of Lines-of-Best-Fit

Notice in Figure 10.1a that the data points tend to move upward as you look from left to right on the graph. This means that as the independent variable *increases,* the dependent variable also *increases.* Contrast this graph with Figure 10.1b where the reverse is true. As the independent variable *increases,* the dependent variable *decreases.* Other patterns of data are shown in Figures 10.1c and d. In these graphs you see one set of data that increases to a peak before decreasing, and in the other set of data where the data starts high but decreases rapidly followed by an equally rapid increase. Sometimes there does not appear to be any pattern in the points on a graph (see Figure 10.1e).

Activity 10.1 Drawing a Best-Fit Line

The rules for constructing a best-fit line for a set of data points on a graph are:

1. The line should be a straight line or a smooth curve.
2. All points should lie either on the line or very near to the line.
3. There should be approximately an equal number of data points on either side of the line.

Shown below are several graphs with lines drawn through the data points. You are to decide whether it is a best-fit line. If you decide the line is not the "best-fit" line, check the reason. The answers can be found in the Self-Check that follows.

How Does the Speed of a Car Affect Gasoline Consumption?

1.
- ❏ A. Line is a best-fit
- ❏ B. Should be curved
- ❏ C. Should be straight
- ❏ D. Too many data points on one side
- ❏ E. Curve is not smooth

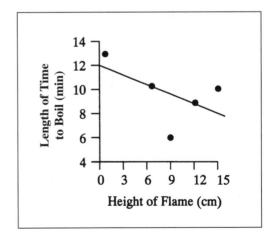

How Does the Height of a Flame Affect Boiling Time?

2.
- ❏ A. Line is a best-fit
- ❏ B. Should be curved
- ❏ C. Should be straight
- ❏ D. Too many data points on one side
- ❏ E. Curve is not smooth

How Does the Time of Day Affect the Amount of Oxygen a Plant Releases?

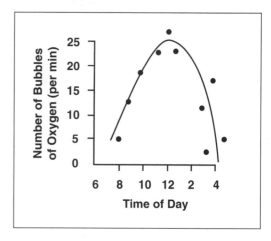

3.
❑   A.   Line is a best-fit
❑   B.   Should be curved
❑   C.   Should be straight
❑   D.   Too many data points on one side
❑   E.   Curve is not smooth

How Is the Height of a Tree Related to Its Age?

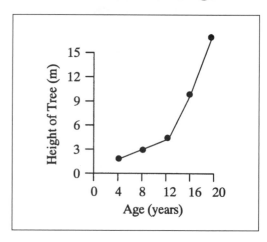

4.
❑   A.   Line is a best-fit
❑   B.   Should be curved
❑   C.   Should be straight
❑   D.   Too many data points on one side
❑   E.   Curve is not smooth

How Does the Time of Day Affect the Temperature of Air?

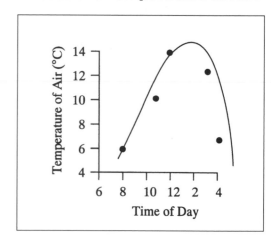

5.
❑   A.   Line is a best-fit
❑   B.   Should be curved
❑   C.   Should be straight
❑   D.   Too many data points on one side
❑   E.   Curve is not smooth

Now try drawing best-fit lines for the graphs below. First, decide whether a straight or curved line fits the points best. Then draw the line. Try to make your lines *average out* the points on the graph. A good line-of-best-fit will usually pass through a few points, be above others, and below still others. Such a line then represents the numerical average of the data points on the line and on either side of the line. As you practice drawing best-fit lines, keep in mind that each data point represents the numerical relationship between the variables being investigated. The best-fit line provides a *picture* of the relationship of all the data points to each other.

6. Draw best-fit lines for these points.

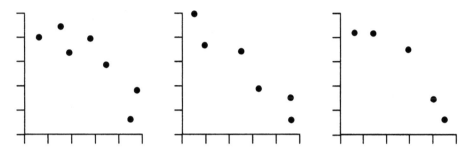

7. Now try a few more. Draw best-fit lines for the points on these graphs.

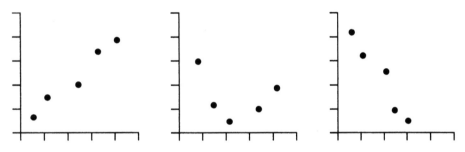

Hint: Holding a graph at arm's length for a moment may help you visualize the general shape of an appropriate best-fit line.

Self-Check _____ Activity 10.1

1. **D,** too many data points on one side. The line should be moved a little toward the upper right. You may have checked **B** also. A curved line could be used here.
2. **B,** should be curved. Some of the points are a long way from a straight line. A curved line drawn in the shape of a "U" would fit better.
3. **A,** line is a best-fit. Notice that the line seems to *average* the points. Some are above the line and some are below.
4. **E,** curve is not smooth. A best-fit line never connects point to point in straight line segments. If it curves, it should be a smooth curve. For the graph in #4, a smooth curve shaped like a "J" would probably be the line of best-fit.
5. **D,** too many data points on one side. There are more points on the inside of the curve than on the outside. The curved line should probably be lowered a little in order to average the points on the graph.

6. Here are the lines we drew. There is some room for differences but your lines should be similar to these.

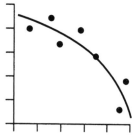 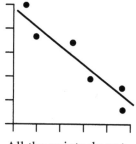 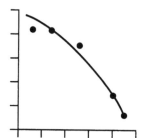

Some of the points are quite far from the line, but equal numbers are found on either side of the line.

All the points do not have to be equal distances from the line.

A straight line would have resulted in some points being a great distance from the line.

7. Be careful about *averaging* the points. The number of points not on the line and their distance from the line should be approximately equal along both sides of the line.

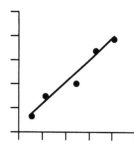 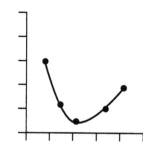 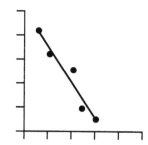

 # Activity 10.2 Writing a Statement of Relationship

Does the Distance of a Race Affect the Number of Runners?

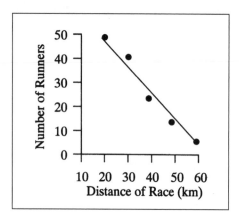

You have just learned how to draw a line-of-best-fit. Now all you have to learn is to write a statement that describes the relationship between the variables on a graph.

For example, if you were given the graph shown here, you should be able to write a statement that summarizes the relationship between the independent (manipulated) and dependent (responding) variables: *As the distance of the race increased, the number of runners decreased steadily.*

One procedure for describing the relationship between variables on a graph uses the following format:

As the _____ _____,

      (independent variable)   *(describe how it was changed)*

the _____ _____.

    (dependent variable)   *(describe how it changed)*

Using this format, a statement of relationship between variables might be written: *As the length of time water was heated increased, the temperature increased.*

> *INVESTIGATION:* Ropes of different diameter are tested to see how much they will hold before breaking.

The answers can be found in the Self-Check that follows.

1. Examine the following graph. Follow the line on the graph as it moves from left to right. Does the value of the dependent variable increase or decrease? _____

2. Using the format given above, write "As" followed by the independent variable and how it was changed.

_____

3. Now write the name of the dependent variable and how it changed.

_____

How Does the Diameter of Rope Affect It's Breaking Point?

You have now described the relationship between the two variables on the graph: **As the diameter of the rope increased, the breaking point increased.**

Try another one.

> *INVESTIGATION:* The number of letters recognized on an eye chart was measured to see if it was affected by distance.

4. Note how the data points move toward the right. What happened to the dependent variable?

_____

_____

5. Write "As" and the independent variable and how it was changed.

_____

_____

6. Follow this with the name of the dependent variable and how it changed.

_____

How Does the Distance from a Chart Affect Letter Recognition?

Two graphs are given here. Write a statement of the relationship between the variables for each graph. Remember to use the procedure on page 193.

Does Number of Objects Affect Spring Length?

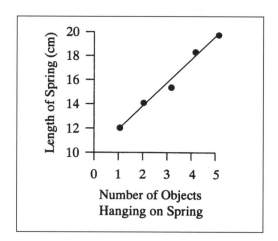

7. _____

   _____

   _____

   _____

   _____

   _____

How Does the Number of People in a Home Affect the Electric Bill?

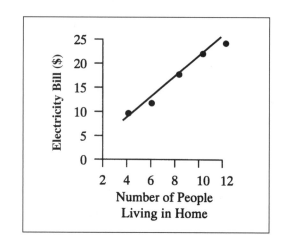

8. _____

   _____

   _____

   _____

   _____

   _____

Self-Check _____ Activity 10.2

1. The values of the dependent variable increased.
2. As the diameter of the rope increased,
3. The breaking point decreased.
4. The values of the dependent variable decreased.
5. As the distance from the chart increased,
6. The number of letters recognized decreased.
   The complete statement of relationship is: *As the distance from the chart increased, the number of letters recognized decreased.*
7. As the number of objects hanging from a spring increased, the length of the spring increased.
8. As the number of people living in the home increased, the electricity bill increased.

The steps for describing the relationship between the variables on a curved line graph are as follows:

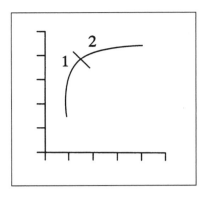

a. Describe the relationship in at least two sentences.
b. First describe the relationship until the curve changes direction.
c. Then tell what the relationship is for the rest of the graph or until another change occurs.

Examine the curve at the right. The first sentence in the description should describe the section marked "1." The second sentence should describe the section marked "2."

A change in the direction of a line indicates a change in the relationships between the variables.

Examine the graph given below.

> *INVESTIGATION:* An ice cube is placed in a glass of water and the temperature of the water is measured every few minutes.

Answers can be found in the Self-Check that follows.

9. Place a mark across the best-fit line about where it first bends.

10. In one sentence describe what happens on the graph up until the mark you drew. _____

_____

11. Now describe what happens on the graph above the mark you drew. _____

_____

How Does the Length of Time an Ice Cube Is in Water Affect the Temperature of the Water?

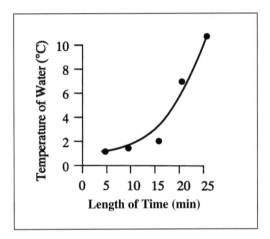

Together the two sentences in #10 and #11 should describe the entire graph. Now try another.

> *INVESTIGATION:* Tomato plants were grown at several different temperatures. The mean number of tomatoes produced by each plant was calculated.

12. Mark where the best-fit line changes direction.

13. Describe what happens on the graph up until the mark you drew. _____

_____

14. Now describe what happens on the graph after the mark you drew. _____

_____

How Does Temperature Affect the Number of Tomatoes Produced?

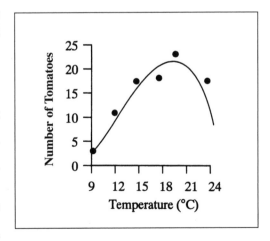

How Does Heating Time Affect the Temperature of Water?

INVESTIGATION: A pan of water is heated over a burner and the temperature is recorded every two minutes.

15. Write a statement of the relationship between the variables shown on the graph.

_____

_____

_____

_____

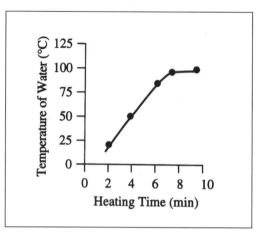

Self-Check

Activity 10.2

9. **How Does the Length of Time an Ice Cube is in Water Affect the Temperature of the Water?**

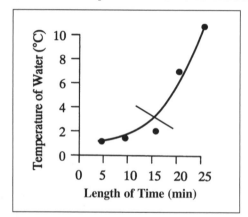

10. *As time passed, the temperature of the water increased slowly.* Your statement does not have to be exactly the same as this but it should be similar.

11. *After 15 minutes the temperature of the water increased rapidly.*

12. **How Does Temperature Affect the Number of Tomatoes Produced?**

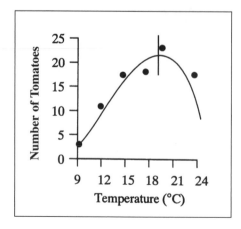

13. As the temperature increased, the number of tomatoes produced increased rapidly.

14. Above a temperature of 19 °C, the number of tomatoes produced decreased rapidly.

15. As water was heated up to 100 °C, the temperature steadily increased. After that, the temperature stayed about the same even though heating continued.

✔

In the materials just completed you learned to draw a best-fit line and write a statement of the relationship between the variables on a graph. In Chapter 9 you learned to construct a graph. Now you are ready to try practice problems in which you put all of these skills together. However, before you do it on your own you will critique two problems in which someone else has made a graph and described it.

Activity 10.3 Analyzing Graphs and Statements That Summarize Trends in Data

Two descriptions of investigations and the data collected for each are given below. Also provided are:

- A graph of the data
- A best-fit line
- A statement of the relationship between the variables

For each investigation, analyze the graph and the statement provided and indicate *what is correct and incorrect about each.* Answers can be found in the Self-Check that follows.

Has the Number of Sea Otters Changed over Time?

INVESTIGATION: Sea otters in a sheltered lagoon were counted over a number of years. These are the recorded data.

Year	Number of Sea Otters
1962	46
1972	42
1982	35
1992	30
2002	26

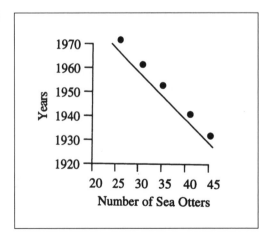

Statement of relationship: Since 1962, the number of sea otters in this location has been steadily decreasing.

1. Which of the following apply to the graph and summary statement on the previous page?

❏ Variables are on wrong axis
❏ Data points are correct
❏ Numerical scale is wrong
❏ Number pairs are in wrong position

❏ Statement is correct
❏ Should be two sentences
❏ Does not include both variables

❏ Best-fit line is correct
❏ Wrong shape line
❏ Line does not average points

Does Distance Affect Accuracy?

INVESTIGATION: The mean number of hits on a target in an archery contest was measured at different distances from the target.

Distance from Target (m)	Mean Number of Hits
15	23
35	22
50	20
75	15
90	4

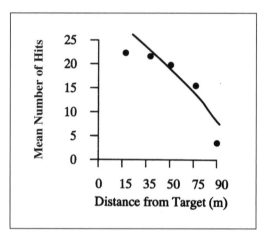

As the distance from the target increased, the mean number of hits decreased steadily.

2.
❏ Variables are on wrong axis
❏ Data points are correct
❏ Numerical scale is wrong
❏ Number pairs are in wrong position

❏ Statement is correct
❏ Should be two sentences
❏ Does not include both variables

❏ Best-fit line is correct
❏ Wrong shape line
❏ Line does not average points

Self-Check _____ Activity 10.3

1. Variables are on the wrong axis.
   Statement is correct.
   Line does not average points (all the points are above the line).
2. Numerical scale is wrong (intervals on horizontal axis are not of equal value).
   Should be two sentences in the description. (As the target distance increased, the mean number of hits decreased slightly. When the target distance was greater than 50 meters, there was a rapid decrease in the number of hits.)
   Wrong shape best-fit line (should be a curved line.)

Activity 10.4 Applying All Your Graphing Skills

Now that you have critiqued some graphs, you should be ready to use the graphing skills you have learned.
  Two descriptions of investigations and a table of data for each are presented. For each, do the following:

a. Construct a graph of the data.
b. Draw a best-fit line.
c. Write a statement of the relationship between the variables.

Answers can be found in the Self-Check that follows.

1. *INVESTIGATION:* A potato is cut in two and allowed to dry in the sun. The mass of the potato is measured as the days pass.

Does Time in the Sun Affect a Potato's Mass?

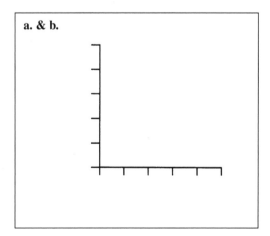

Time in the Sun (days)	Mass of Potato (grams)
1	330
7	300
12	180
14	150
21	160
26	120

c. _____

_____

_____

2. *INVESTIGATION:* A fire chief did an analysis of his men at work. He measured the mean time it takes a fireman to climb ladders of different lengths.

Does the Length of a Ladder Affect the Time It Takes to Climb It?

Length of Ladder (m)	Time to Climb (sec)
1	2
2	5
3	8
8	18
12	22
15	53

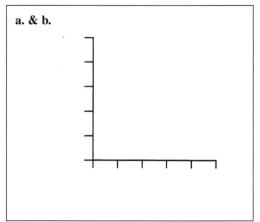

a. & b.

c. _____

_____

_____

Self-Check _____ Activity 10.4 ✔

1. As the days in the sun increases, the mass of the potato drops rapidly until the 15th day. After that, the loss in mass is very slow and seems to be stopping.

Does Time in the Sun Affect a Potato's Mass?

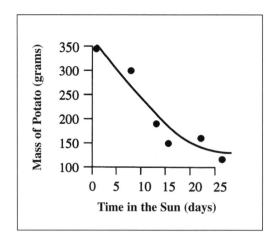

2. Note: Either a curved or straight best-fit line could be used here. As the length of the ladder increased, the time to climb it steadily increased. If you drew a curved line-of-best-fit, you may want to add a second sentence like this one: Above lengths of 12 meters, the time to climb increased more rapidly. Acquiring more data might help to determine the shape of the line-of-best-fit.

Does the Length of a Ladder Affect the Time It Takes to Climb It?

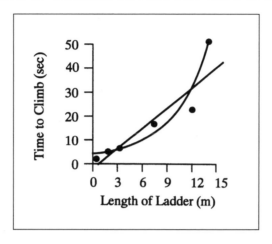

If you were successful on the last two graphs, you have learned some important skills that are helpful in solving science problems. In the next chapter you will learn to acquire and process your own data.

Now take the Self-Assessment for Chapter 10.

Self-Assessment

Describing Relationships between Variables

1. A description of an investigation and a table of data are given here.
   a. Construct a graph.
   b. Draw a best-fit line.
   c. Write a statement of the relationship between the variables.

   INVESTIGATION: An investigation was carried out to determine the relationship between the size of an automobile motor and the gasoline mileage.

Size of Motor (horsepower)	Kilometers per Liter of Gasoline
47	7.0
100	5.0
140	4.0
193	3.5
227	3.0

a. & b.

c. _____
_____
_____
_____

2. Draw a best-fit line for the points on the graph.

Does Adding Nitrogen Affect Corn Production?

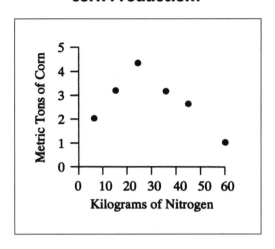

3. Write a statement of the relationship between the variables shown on this graph.

   INVESTIGATION: A weather station kept a record for a ten year period of the mean amount of rainfall during several months of the year.

   _____

   _____

   _____

How Does Month of the Year Affect Rainfall?

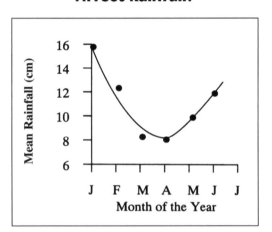

4. Write a statement of the relationship between the variables on this graph.

   INVESTIGATION: Some soldiers were tested to see if the number of kilometers they could hike in an hour was affected by the temperature.

   _____

   _____

   _____

How Does Air Temperature Affect Hiking Distance?

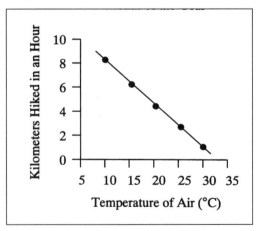

Self-Assessment Answers

1.

Does Size of a Motor Affect Gasoline Consumption?

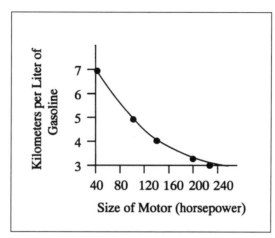

As the size of the motor increased, the number of kilometers per liter of gasoline decreased. However, the decrease was slower for motors above 120 horsepower.

2.

Does Adding Nitrogen Affect Corn Production?

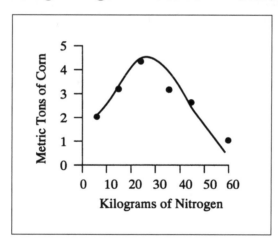

3. Between January and April, the mean rainfall per month steadily decreased. The mean rainfall per month steadily increased from April to June.

4. As the temperature of the air increased, the number of kilometers traveled by soldiers steadily decreased.

High Stakes Testing

A sample multiple-choice item from State Standardized Exams.

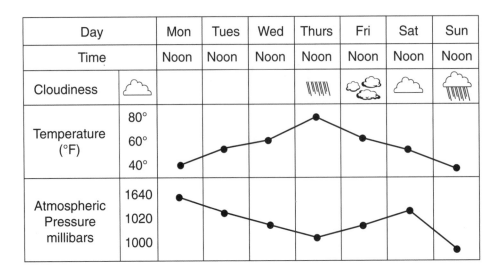

Day		Mon	Tues	Wed	Thurs	Fri	Sat	Sun
Time		Noon	Noon	Noon	Noon	Noon	Noon	Noon
Cloudiness								
Temperature (°F)	80° 60° 40°							
Atmospheric Pressure millibars	1640 1020 1000							

Look at the diagram for the days Monday through Thursday. Choose the best description of the relationship between temperature and pressure for those days.

  A.   *As the temperature rose, the pressure remained the same.*
  B.   *As the pressure rose, the temperature remained the same.*
  C.   *As the pressure rose, the temperature dropped.*
  D.   *As the temperature rose, the pressure dropped.*

Illinois: ISAT: Grade 4 Sample Test.

http://www.sciencenetlinks.com/Lessons.cfm?DocID=146
www.kidport.com/Grade4/TAL/G4-TAL-Relationships.htm

WEBSITES FOR ACTIVITIES

A Model for Assessing Student Learning

Assessment Type: Performance Task

Preparation

Pre-assemble a small plastic bag of materials for each student or group of students doing the task at the same time. Use large washers, small fishing sinkers, or similar objects as the pendulum bob. Make 6 strings, each with a loop tied at one end and a paper clip at the other end. Open the paper clips slightly to serve as a hook for the pendulum bob. The string lengths from the top of the loop to end of the pendulum bob should be 10, 20, 30, 40, 50, and 60 centimeters. Have several rolls of masking tape available or distribute strips of tape to participating students. *Groups of students may complete Part I of this task together but Part II should be completed individually.*

Directions to the Student

Part I

1. Check your materials.

   ✔ pencil
   ✔ 6 strings with a paper clip tied to one end
   ✔ 1 large washer (or other weight)
   ✔ roll or strip of tape
   ✔ clock or watch

2. Tape your pencil to your desk so it hangs over the edge.

3. Hang the shortest string (10 cm) from the pencil and hook your washer on the paper clip to make a pendulum.

4. Hold the pendulum even with the top of your desk and release it.

5. Count the number of times the pendulum swings back to where you released it in 15 seconds. Record your data in the table.

6. Repeat steps 3–5 using each of the 5 remaining strings (20, 30, 40, 50, and 60 cm).

Length of Pendulum (cm)	Number of Swings Back
10	
20	
30	
40	
50	
50	
60	

Part II

1. Construct a graph of the results of this activity.

2. Draw a best-fit line.

3. Write a statement of the relationship between the variables.

_____

_____

_____

Chapter 11

Acquiring & Processing Your Own Data

National and State Standards Connections

- Science as Inquiry: Use appropriate tools and techniques to gather, analyze, and interpret data. (NSES 5–8)
- Collect data using observations, surveys, and experiments. (NCTM 3–5)
- Conduct multiple trials to test a prediction and draw conclusions about the relationships between predictions and results. (California: STAR, Grade 4)

MATERIALS NEEDED

- ✔ 4 Pyrex beakers (100 mL)
- ✔ 1 graduated cylinder
- ✔ 1 measuring spoon
- ✔ 1 hot plate
- ✔ 1 thermometer
- ✔ 1 timer
- ✔ sugar
- ✔ 1 plastic drinking straw
- ✔ long piece of string

- ✔ 1 balloon
- ✔ masking tape
- ✔ 1 quart size sealable bag
- ✔ 5 small bathroom-size paper cups (about 3 oz.)
- ✔ 1 sharpened pencil
- ✔ 1 container of water
- ✔ 1 clock or watch with a second hand

Purpose

In the previous three chapters you have been working with many data tables. However, the number pairs in these data tables were produced by someone else. In this chapter you will carry out several investigations and produce your own tables of data.

Objectives

After studying this chapter you should be able to:

1. Conduct an investigation and construct a table of data. (See Chapter 8.)
2. Construct a graph of the data and a statement of the relationship between the variables. (See Chapters 9 and 10.)

Measurement Skills

In this chapter it is assumed that you know how to make measurements of mass, length, time, temperature, force and volume. If you are not sure that you know how to make these measurements, you may wish to review Chapter 4.

Graph Titles

You may have noticed that all the graphs in Chapters 9 and 10 have titles. A graph title is very important because it communicates the purpose of the graph to the reader. Two simple formats for writing titles that will be helpful to you and your future students are:

Sentence format: *The Effect of (independent variable) on the (dependent variable).*

Question format: *How Does the (independent variable) affect the (dependent variable).*

For example:

The Effect of a Basketball Player's Weight on the Height He Can Jump

or

How Does the Weight of a Basketball Player Affect the Height He Can Jump?

When you are asked to construct a graph, be sure you give it a title by using either of these title formats. Each time you use a Self-Check also check your graph title.

Activity 11.1 Initiating an Investigation with a Why Question

1. An experiment generally begins with a problem. Someone observes something occurring and wonders *Why?* For example, everyone knows that the old cliche, *a watched pot never boils,* is not true; however, it does raise an interesting problem. What determines the time it takes water to heat? Examine this problem a little closer. What are some of the variables that could affect the heating time of water? Answers can be found in the Self-Check that follows.

_____   _____

_____   _____

_____   _____

If the variable, *amount of dissolved material,* was selected for investigation, one could make this prediction: The more material that is dissolved in water, like sugar or salt, the longer it will take for water to heat. The first exercise of data gathering will be to conduct the experiment proposed to test this hypothesis. The directions you are to follow are given below.

2. ➡**GO TO** the supply area and obtain the following:

4 Pyrex beakers (100 mL)

1 graduated cylinder

1 measuring spoon

1 hot plate

1 thermometer

1 timer

sugar

Label the beakers 0, 1, 2, and 3. Measure 50 mL of water into each. Dissolve one spoon of sugar in beaker 1, 2 spoons in beaker 2, and 3 in beaker 3. Place no sugar in beaker 0. Heat each beaker for three minutes. Record the change in temperature in the table below.

How Does the Amount of Sugar Affect the Change in Water Temperature?

Amount of Sugar (spoons)	Temperature Change (°C)
0	
1	
2	
3	

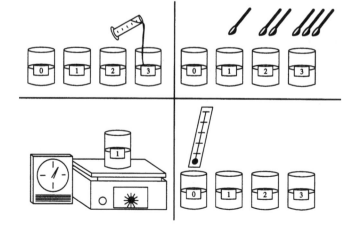

3. Why bother heating plain water where no amount of sugar was added?

_____

_____

_____

_____

4. Using the data obtained from your investigation, construct a graph and write a *statement* about the relationship between the amount of dissolved sugar and the change in temperature. Be sure to give the graph a title.

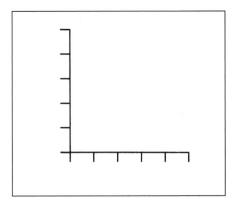

_____

_____

_____

_____

_____

_____

Self-Check _____ Activity 11.1

1. Amount of water   Shape of the pot
   Amount of dissolved material   Type of heat source
   Height above sea level

   You may have thought of some that were different from these, such as the kind of metal pot or the kind of water.

2. **How Does the Amount of Sugar Affect the Change in Water Temperature?**

Amount of Sugar (# of spoons)	Temperature Change (°C)
0	38
1	35
2	32
3	31

These are data we collected.

3. The beaker with no sugar serves as a standard of comparison to which all the other beakers with different amounts of sugar are compared. In order to know whether sugar causes any change, it

is necessary to know what happens to the water temperature when no amount of sugar is used. Most experiments include a standard of comparison, called a **control** or control group. In some experiments, such as this one, the control is called a **no treatment** control. In other experiments all groups receive a treatment (some amount of the independent variable). The experimenter must select one of the levels of the independent variable to serve as the control group. The level selected is usually the normal or typical case. For example, in an experiment on the effect of depth of seed on seed germination, the control might be the recommended or normal planting depth; other planting depths might be deeper or shallower than the control. This kind of control is called an **experimenter-selected** control.

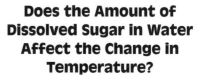

Does the Amount of Dissolved Sugar in Water Affect the Change in Temperature?

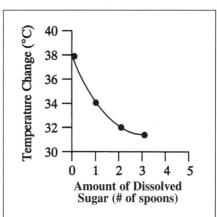

4. **Statement of Relationship**

    The temperature change decreases, as the amount of dissolved sugar increases. The addition of more than two spoons of sugar results in no increase in temperature change.

    Your statement may be different from this one if your graph differs.

    This is a graph of the data collected earlier. Yours may be different because the data you gathered were different from ours.

Here is another practice exercise:

Leticia and her cooperative learning team found an activity that might help them learn about their assigned project, space travel. The activity directions were:

1. Thread a plastic drinking straw onto a string that is long enough to reach from one side of the room to another.
2. Stretch the string tight between two sides of the room.
3. Tape one side of a quart size plastic bag to the straw as illustrated below.
4. Blow up a balloon, hold it shut, and place the balloon in the bag.
5. Release the balloon and watch the balloon travel along the string—like a rocket.

In Activity 11.2 that follows, you will collect data to test the prediction that was made by this team of students.

Activity 11.2 Initiating an Investigation with an Activity and a Question

1. **→GO TO** the supply area and obtain the following materials and set up the activity as Leticia and her team would.

    You will need:

    ✔ 1 plastic drinking straw
    ✔ long piece of string
    ✔ 1 balloon
    ✔ masking tape
    ✔ 1 quart size sealable bag

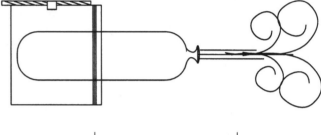

Follow the directions given on page 211 for the activity. Conduct several trials to find out how the "rocket" works.

Leticia and her team wanted to turn this activity into an experiment by changing one variable and measuring the response. They wanted to know if the number of breaths blown into the balloon would affect the number of meters it traveled along the string. Their prediction was: As the number of breaths blown into the balloon are increased, the balloon will travel farther along the string.

Use your balloon rocket to test their prediction. Try different numbers of breaths (at least four different ones) and conduct 3 repeated trials of each number of breaths. Add a title, label the columns, and record your data in the table. Calculate the mean distance that the balloons traveled for each different number of breaths.

	Trials	
1	2	3

The table of data constructed by Leticia's team can be found in the Self-Check that follows. Your data will be different, but your title and column labels should be similar.

Answers to the following can be found in the Self-Check.

2. What other variables might affect how far the rocket travels?

_____

_____

_____

_____

_____

_____

Title:

3. Using the data obtained from your investigation, write a title, construct a graph and summarize the relationship between the number of breaths blown into the balloon rocket and the mean distance the rocket traveled.

Summary Statement:

_____

_____

_____

_____

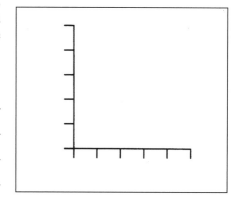

Self-Check

✔

1. ## How Does the Number of Breaths Affect the Distance Traveled by a Balloon Rocket?

Number of Breaths	Distance Traveled (m)			Mean Distance Traveled (min)
	Trials			
	1	2	3	
1	1.7	2.0	2.3	2.0
2	5.4	6.0	5.6	5.6
3	9.3	9.1	8.2	8.9
4	10.5	9.2	10.3	10.0

2. Here are some possible variables about each of the materials used in this activity that might affect how far the balloon rocket travels.

Balloons
Shape
Diameter
Length
Composition
Age
Previous use

Straw
Type
Mass
Length
Composition

String
Tautness
Angle
Length
Composition

You also may have thought about keeping the amount of air in each breath the same in some way, or perhaps, the way the balloon rocket is launched each time.

3. Leticia's team made the following graph. Your data will be different but your axis labels and graph title should be similar. Compare your statement of relationship with the one the team wrote.

Statement of Relationship

The distance the balloon traveled increased steadily as the number of breaths blown into the balloon increased. (Note that it would be helpful to have collected more data.)

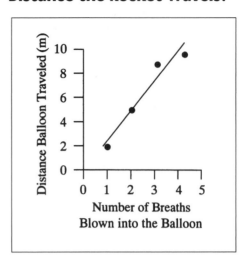

Does the Number of Breaths Blown into a Balloon Rocket Affect the Distance the Rocket Travels?

Activity 11.3   Another Question and an Activity to Answer It

Why do you suppose flower pots have holes in the bottom? Have you ever noticed that they always seem to have just one hole? If the hole is there for drainage, would two holes drain faster? What about three holes? *How does the number of drainage holes in a container affect the time for it to drain?* How much difference do you think increasing the number of holes will make?

➡**GO TO** the supply area and gather the following materials:

- 5 small bathroom-size paper cups (about 3 ounces)
- 1 sharpened pencil to make holes
- 1 container of water to fill the cups
- 1 container to catch the draining water, or use the sink
- 1 clock or watch with a second hand

1. Use the sharpened pencil to make a hole in the bottom of the cup. Push the pencil all the way through the hole to make the largest diameter hole possible.
2. Place your finger over the hole and fill the cup with water.
3. Hold the cup over a sink or container.
4. Remove your finger and time how long it takes the cup to drain.
5. Record the time in your data table and conduct repeated trials if you have time.
6. Repeat the procedure by making 2, 3, 4, and then 5 holes in a new cup each time.
7. If you conducted repeated trials for each number of holes, record all your data in the data table on the next page. Calculate the mean drainage time for each number of holes.
8. Construct a graph and write a statement of the relationship between the variables from the data obtained. Remember to add a title to both your data table and graph. The answers for this activity can be found in the Self-Check that follows.

Title:

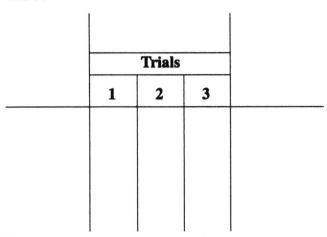

	Trials		
	1	**2**	**3**

Title:

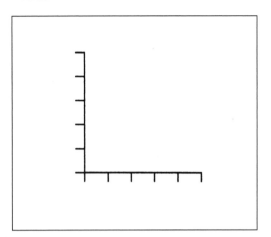

Summary Statement:

_____

_____

_____

_____

Self-Check _____ Activity 11.3

These are the data gathered during the investigation when we did it. You can compare your results with ours. Remember that the data you obtain and the graph you make may differ from ours and yet be correct.

How Does Number of Drainage Holes Affect the Time to Drain?

Number of Drainage Holes	Time to Drain (sec)			Mean Time to Drain (sec)
	Trials			
	1	**2**	**3**	
1	13	12	13	13
2	5	6	4	5
3	4	3	4	4
4	3	4	3	3
5	3	3	3	3

How Does Number of Drainage Holes Affect the Time to Drain?

As the number of drainage holes was increased from 1 to 3, the time required to drain the cup decreased dramatically. Thereafter, the time appeared to level off as more drainage holes were added.

You now have some experience in collecting, tabulating, and graphing data when you were provided with a problem to investigate. In the following chapters you will examine additional parts of an experiment to ultimately help you design your own investigation in the last chapter.

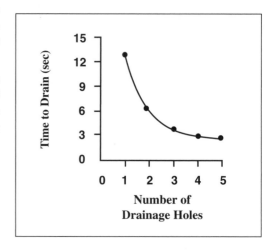

 **Self-Assessment**

Acquiring and Processing Your Own Data

How does amount of exercise affect a person's pulse rate? To answer this question about yourself, conduct the following investigation.

Sit quietly for 5 minutes and then count your number of heartbeats for 15 seconds. Now quickly step up onto a stool or stair step 5 times. Count your heartbeat for 15 seconds. Rest until your pulse rate returns to the resting rate. Quickly step up onto the stool 10 times. Count your heartbeat for 15 seconds. Repeat this procedure for 15, 20, and 25 step-ups. If you had time, you could improve the validity of your results if you conducted repeated trials of each level of the independent variable, and then found the *mean* number of heartbeats for each amount of step-ups.

Construct a data table, a graph, and write a statement of the relationship between the variables for the data you gather.

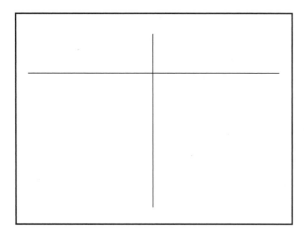

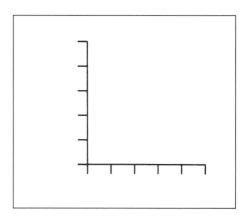

Summary Statement:

_____

_____

Self-Assessment Answers

Here are the data gathered from our investigation. Your data will almost certainly be different.

Does the Number of Step-Ups Affect a Person's Heart Rate?

Numbers of Step-ups	Number of Heartbeats (in 15 seconds)
0	20
5	26
10	28
15	32
20	36
25	39

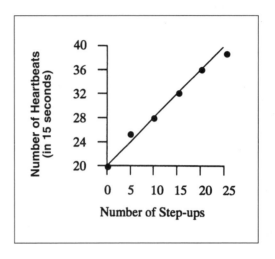

Summary Statement: The pulse rate increases as the number of step-ups increase.

High Stakes Testing

A sample multiple-choice item from State Standardized Exams.

Temperature of a Liquid in Degrees Celsius (°C)

Minute 1	Minute 2	Minute 3	Minute 4	Minute 5	Minute 6
21 °C	23 °C	25 °C	27 °C	29 °C	?

Two students did a science experiment in class. They recorded the temperature of a liquid each minute. The chart above shows their data. What will be the temperature of the liquid at the 6-minute reading?

A. 29 °C
B. 31 °C
C. 32 °C
D. 33 °C

Kentucky Core Content Test Grade 4, Release Items Spring 1999.

askeric.org/Virtual/Lessons/Mathematics/Process_Skills/MPS0016.html
askeric.org/Virtual/Lessons/Mathematics/Probability/PRB0005.html
http://www.hhmi.org/coolscience/airjunk/index.html
sln.fi.edu/fellows/fellow7/mar99//probability/index.shtml

WEBSITES FOR ACTIVITIES

A Model for Assessing Student Learning

Assessment Type: Technology-Based Assessment

Learner Profile™ is a set of software assessment tools and an example of how technology is being applied to classroom instruction, assessment, and record keeping. Teachers can use a desktop computer or a Palm™or other OS compatible personal digital assistant (PDA) to systematically evaluate students. By transferring student names and learning objectives to a PDA, teachers can use the stylus to instantly record observations of student behavior in any educational setting—the classroom, the science laboratory, school grounds, and field trips. Learning indicators are added to the database as measures of objectives and standards. These are distinct observable actions performed by students that demonstrate an achievement in learning, such as local, state, and national standards. The observable actions are assessed using qualifiers that describe the learner's level of competency.

For ultimate mobility in the classroom, a teacher can use a PDA to record classroom observations as they occur, take attendance, check homework, or grade tests. As students read, calculate, perform, and demonstrate their learning, teachers select their names, the observable learning indicators, and the appropriate qualifiers to record their level of mastery. At the end of the day, teachers send their observations back to their desktops. Reports to parents, students, colleagues, and administrators can be easily created with the assessment information collected.

Reviewing and Deleting Observations

Using the Observation Maintenance feature, you can browse the observations you have recorded and delete observations you may have recorded in error.

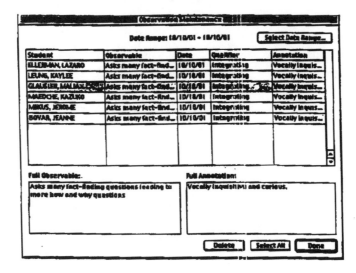

Chapter **12**

Analyzing Investigations

National and State Standards Connections

- Develop general abilities, such as . . . identifying and controlling variables. (NSES 5–8)
- Select and use various types of reasoning and methods of proof. (NCTM Pre-K–2)
- Identify the single independent, the dependent and the controlled variables in an investigation. (California Content Standards, Grade 5)

Purpose

Before you can design your own investigations, you need to learn to recognize the parts of a typical investigation. What are the variables under study? What hypothesis is being tested? These and other questions can be answered by analyzing an investigation.

Objectives

After studying this chapter you should be able to:

1. Identify the independent and dependent variables and the constants in an experiment.
2. Identify the hypothesis being tested when supplied with a description of an investigation.

APPROXIMATE TIME FOR COMPLETION — 30

You're hungry and want boiling water for spaghetti fast. Should you put the salt in before the water boils or after? What effect does adding salt to water have on the time it takes the water to boil?

Suppose you wanted to test the hypothesis, *If the amount of salt added to water increases, then the time it takes to boil will also increase.* To test this idea, you could heat three pots of water. Add a small amount of salt to one pot and a larger amount to another. Heat the third pot with no salt added to serve as a standard of comparison, or control. You would need to keep several other factors the same, such as the amount of water in each pot and the kind of metal each pot was made of. Without making these and others factors your constants, you would not be sure if the results were the result of differing amounts of salt or some other variable that was not kept the same throughout the experiment.

In an experiment we want to be able to say that the independent variable and only the independent variable affects the dependent variable. We must make sure that all other variables that could affect the results are prevented from having an effect.

The term **variable** describes factors that change or could possibly change in an experiment. A variable that might affect an experiment but is kept from doing so is called a **constant.** Constants are all those factors that are kept the same so they are prevented from affecting the outcome of the experiment. Some people refer to an experiment with explicit constants as a controlled experiment.

 # Activity 12.1 Holding Factors Constant

What factors were kept the same in this experiment on the effect of amount of coffee on the color of coffee produced?

Six cups of tap water were placed in each of four identical coffee makers. Then 1 scoop of Brand X coffee was added to the first coffee maker, 2 scoops to the second, 3 to the third, and 4 to the fourth coffee maker. The coffee maker ran through its full cycle each time.

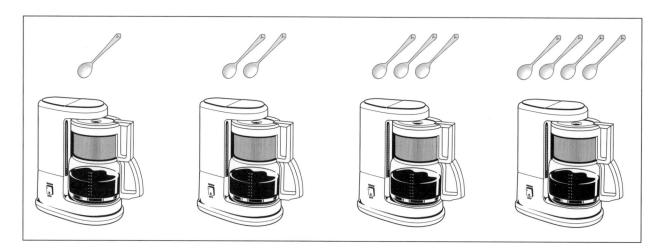

Answers can be found in the Self-Check that follows.

1. State the factors that were held constant and describe how they were kept the same.

   _____   _____

   _____   _____

   _____   _____

2. *What factors are held constant in the following experiment?*

   A herd of Angora goats is divided into two groups. Both groups are housed in the same building, fed at the same time each day, and given the same amount of water. One group gets Brand X feed and the other Brand Y.

   _____   _____

   _____   _____

   _____   _____

3. What additional factors did you assume were kept the same in the previous experiment, even though they were not specifically mentioned in the description?

   _____   _____

   _____   _____

   _____   _____

4. From this drawing alone, what factors would you infer were kept the same in this experiment setup?

   _____

   _____

   _____

   _____

5. Consider an experiment to test the prediction that the more light plants receive, the taller the plants will grow. Amount of light, the independent variable, changes because you purposely changed the light. The other variable, height of the plants, is your dependent variable and it changes in response to how you changed the independent variable. All other factors must be kept the same. Among the constants in this experiment are:

   - All the plants are the *same* size
   - The *same* soil in each pot
   - Water at the *same* time each day
   - Given the *same* amount of water
   - Kept in the *same* place

   Suppose two groups of plants were used. One group of plants would receive more light than the other group of plants. All other potential variables would be kept the same for each group of plants. They would become the constants for this experiment.

   By keeping all other factors the same, the experiment is a fair test of how amount of light affects the height of plants.

   Suppose in this experiment one group of plants was grown at 16 °C.

   At what temperature would you keep the other group of plants? _____

6. What factors should be held constant to test the following hypothesis?

If the amount of salt added to ice is increased, then the temperature of the mixture will decrease.

_____   _____

_____   _____

_____   _____

Self-Check _____ Activity 12.1 ✔

1. **Constants**   **How they were kept the same**

Constants	How they were kept the same
Amount of water	same amount—6 cups each time
Type of coffee maker	identical coffee makers were used
Kind of water	same kind—tap water
Type of coffee	same brand of coffee—brand X

You may have listed other constants besides these, such as the size of the scoop. Although the size of the scoop was not explicitly stated, we can hope that the same scoop was used each time. The independent variable in this experiment is the amount of coffee and the dependent variable is the resulting color of the coffee. Any differences in color that we observe we can attribute to the independent variable, amount of coffee, provided we kept all other factors constant.

2. Type of goat   Feeding time
   Type of housing   Amount of water

If you listed constants other than these, you may have included other factors that need to be held constant but were *not explicitly* stated in the experiment description.

3. Number in each group   Age of the animals
   Temperature of housing   Bedding characteristics
   Amount of feed available

This time you may have thought of others. This is all right if you believe that they could affect the dependent variable.

4. Amount of liquid   Kind of liquid
   Size of container   Temperature of liquid
   Shape of container   Pressure on liquid surface

By this time you are probably saying, *How in the world do they expect me to get all of those? Besides, I thought of some others.* This is just the point. In any experiment there could be many potential variables; so many that it is often difficult to think of them all. A good way to help you identify these potential variables is to make a list of the materials and environmental conditions used in an experiment. Then, think of ways to vary each of the materials and conditions that might affect the outcome of the experiment.

5. 16 °C

6. *Several identical containers* were filled with the *same amount* of ice. Use the *same kind of salt* in each and measure the temperature with the *same kind of thermometer* in each.

You may have thought of others. As long as you were attempting to keep everything in all setups the same except the amount of salt, it would be appropriate.

✔

Activity 12.2 Constructing Well-Stated Hypotheses

Scientists are interested in explaining events. To do so they conduct investigations to determine how independent variables affect dependent variables. In order to plan what investigation should be conducted a statement called a hypothesis is made. A hypothesis is a prediction about the effect an independent variable will have on a dependent variable.

Answers can be found in the Self-Check that follows.

1. What two variables are included in a hypothesis?

   _____

   _____

A hypothesis can be written in several ways. One typical way is in the form of **"If . . . , then . . . "** sentences.

If <u>the amount of salt added to ice</u>   <u>is increased,</u>   **then**   <u>the temperature of the mixture</u>   will   <u>decrease.</u>
   (independent variable)   (describe how you change it)   (dependent variable)   (describe the effect)

The same hypothesis could also be written in other formats as well:

- The greater the amount of salt added to ice, the lower the temperature of the mixture.
- Increased amounts of salt added to ice causes lower temperatures of the mixture.
- As the amount of salt added to ice increases, the temperature of the mixture decreases.

Initially, young students benefit from some structure, so you might find it useful to give them a format for writing hypotheses to get them started, which is similar to the formats for titles suggested in Chapter 11.

If . . ., then . . . format:

If the _____   _____ ,
           (independent variable)   (describe how it will be changed)

then the _____ *will* _____.
             (dependent variable)   (describe the effect)

2. A hypothesis attempts to predict an outcome. *Which of the following predicts an outcome?*
   - ❏ 1. As more salt is dissolved in water, the water will become cloudy.
   - ❏ 2. The earth's crust contains 90 elements.
   - ❏ 3. Magnetism and gravity are not the same.
   - ❏ 4. If the length of a vibrating string is increased, the sound will become louder.

Young students often think of a prediction as a guess, even a *wild* guess. Some teachers introduce the term *educated guess* to help students understand that predictions are based on previous related knowledge and/or experience and are therefore not wild, random guesses. However, many students find it difficult to see the distinction among guesses of any type. You may find it more useful to ask students to write their hypothesis and then add an additional statement that explains *why* they predicted the way they did. Here are a few examples of what students might write:

    a. As the temperature of a cold-blooded animal's environment increases, the temperature of its body also increases. (Cold-blooded animals can not sweat to cool themselves, or maybe because they have no fur or fat to insulate their bodies.)

    b. A change in weather causes a change in mood. (My grandmother who has arthritis complains more on rainy days.)

3. Which of these is stated as a hypothesis?
   - ❑ 1. If cloud cover serves as an insulator, then the surface temperature of the earth should get colder on cloudless nights.
   - ❑ 2. Green leaves manufacture food, stems transfer food, and roots store the food.

4. Which of these are hypotheses?
   - ❑ 1. The colder the temperature, the slower plants grow.
   - ❑ 2. The deeper one dives, the greater the pressure.
   - ❑ 3. Algae are living organisms.

5. Now you try to write a hypothesis. Write a statement that predicts the outcome if the *amount of light* is one variable and the other is *plant growth.*

    _____

    _____

You now have had some practice in identifying variables and hypotheses when given partial descriptions of investigations. Next you will analyze the entire investigation and identify the variables involved and the hypothesis being tested.

6. Carlos was interested in determining the effect of the number of seeds planted in a particular space on the growth of the plants. He planted the same type of radish seeds 1 cm deep in identical milk cartons using the same kind of potting mix. In the first carton he planted 5 seeds, in the second 10 seeds, in the third 15 seeds, and in the fourth carton he planted 20 seeds. Each milk carton received the same amount of water twice a week. He measured the length of the leaves at the end of 3 weeks.

    What factors were kept the same (the constants)?

    _____

    _____

    What variable was manipulated (the independent variable)? _____

    Which variable was expected to respond (the dependent variable)? _____

    What was the hypothesis being tested? _____

    _____

7. Here is a description of another experiment:

    Is there a relationship between the amount of training received and the length of time a learned behavior persists in insects? Select a number of sowbugs (pill bugs or 'rollie-pollies') which always turn right when entering the intersection of a T-shaped maze. Using the tendency of sowbugs to avoid light, it is possible to train them to turn left by shining a strong-light from the right as they enter the intersection. Subject an animal to 1, 5, 10, 15, or 20 training sessions. Test each animal once an hour by running it through the T-maze.

What were some of the constants?

_____   _____

_____   _____

What was the independent variable? _____

Which was the dependent variable? _____

8. Suppose you read that farmers in Florida, where the soil is sandy, had to irrigate their crops frequently, but that farmers in Virginia, where the soil often has a high clay content, had to irrigate their crops less frequently even in the heat of summer. You wonder, "Will different kinds of soil hold different amounts of water?" The drawings below indicate the kinds of materials available to you. Think about how you might conduct an experiment to answer your question.

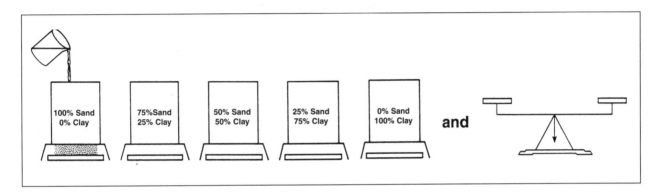

What would be your independent variable? _____

Which would be your dependent variable? _____

What would be some constants in your experiment? _____

_____

What hypothesis would you test? _____

_____

Self-Check _____ Activity 12.2

1. Independent variable
   Dependent variable

2. #1 and #4 predict the effect one variable will have on another.

3. In #1 the effect of the independent variable (amount of cloud cover) on the dependent variable (surface temperature at night) is predicted. Therefore, #1 is a hypothesis. To justify their hypothesis, a student might add, "I read that clouds act as insulators."

In # 2, only factual information is given, and is therefore not a hypothesis.

4. 1, 2

#3 is merely a statement of fact and is therefore not a hypothesis.

1 and # 2 are hypotheses. As an additional step, students could have added the following to justify their hypotheses:

#1 (I know that greenhouses are kept very warm.)

#2 (The deeper you go, the more water there is pushing down on you.)

One advantage to having students write a justification statement for their hypotheses is that it gives you, their teacher, some insight into their thinking. Their reasons may not always be correct, but you will at least know why they hypothesized as they did.

5. The greater the amount of light, the greater the amount of plant growth.

or

The less light the plant receives, the less the plant will grow.

These are just some of the predictions that you could have written as you thought about what might happen to one variable as you purposely changed another. Information you know and your prior experiences both contribute to your ability to hypothesize about what will happen. If you test your hypothesis, you are testing your reasoning as well. When designing an experiment to test your hypothesis, you should use only one independent variable. You could, if you wanted, have more than one dependent variable because there are often multiple ways to measure the resulting effects. You would also want to be sure that all other potential variables are kept the same in your experiment. These factors become your constants.

6. Some factors that were kept the same are: kind of seed, planting depth, soil, environmental temperature, amount of water, kind of container, and amount of light received. Because the number of plants in a particular space was manipulated and the length of the leaves was expected to respond, the hypothesis probably was: *As the number of plants in an area increases, the length of the leaves will become shorter.* Or it could have predicted that the leaves would be longer. In stating a hypothesis, the decision as to what the effect will be can be based on past experience, related information, and even a hunch, but should never be a wild guess. Until data are gathered and interpreted, however, one prediction may be just as valid as another.

7. Type of animal   Strength of light source
   Shape of maze   Environmental temperature

These are just some of the constants. You may have stated others.

The independent variable was the amount of training, while the dependent variable was the length of time a learned behavior persisted.

8. The percent of sand and clay is the independent variable, while the mass of the retained water is the likely dependent variable. Factors that could vary, such as the amount of soil, volume of the containers, soil temperature, kind of sand, and kind of clay must be kept the same. The hypothesis most likely being tested is: *As the amount of clay in the soil increases, the amount of water retained by soil also increases.*

You now have had some practice in analyzing investigations, looking at the variables involved, and identifying the hypothesis being tested. In the next chapters you will begin the task of designing your own investigation.

Now take the Self-Assessment for Chapter 12.

 # Self-Assessment   Analyzing Investigations

Read the description of this investigation and then answer the questions below.

1. A study was conducted to determine how the number of paper clips picked up was related to the number of dry cells connected to the electromagnet. The magnet, connected to 1, 2, 3, 4 or 5 D cells, was placed on the top of a pile of 100 paper clips and lifted.

    a. Identify the constants in the above investigation.

    _____   _____

    _____   _____

    b. Identify the independent variable. _____

    c. Identify the dependent variable. _____

    d. State the hypothesis being tested. _____

    _____

2. Check the statements below that are stated as a hypothesis.

    ❏ a. Baking powder is used in biscuits.
    ❏ b. The brighter the color of an orange, the juicier the fruit.
    ❏ c. Brass contains copper and zinc.
    ❏ d. As the amount of antifreeze increases, the temperature at which the mixture freezes gets lower.

3. Check the statements that are hypotheses:

    ❏ a. As the amount of cabbage in a soup increases, the intensity of the odor also increases.
    ❏ b. Most apples are red, but some are yellow.
    ❏ c. The faster a river flows, the greater the erosion.
    ❏ d. Dental floss is waxed so that it slips between teeth.

4. Suppose you wished to test the hypothesis stated below. Which of the factors listed should be kept the same in the experiment? *Hypothesis: The warmer the water, the faster an aspirin will dissolve.*

    ❏ a. amount of water
    ❏ b. brand of aspirin
    ❏ c. temperature of water
    ❏ d. size of the tablet

5. Write a justification statement for the following hypothesis:

    Hypothesis: The thicker the coating of peanut butter, the slower a sandwich will be eaten.

    _____

    _____

Self-Assessment Answers

1. a. number of paper clips in the pile, type of dry cell, size of the electromagnet, shape of the pile
   b. number of dry cells
   c. number of paper clips picked up
   d. The more dry cells connected to an electromagnet, the more paper clips it will pick up. If you added a justification statement, you might have said, *"Adding more batteries (dry cells) will add more energy."*
2. b and d
3. a and c
   For "a," students who wrote a justification statement might have written that *"Cooked cabbage smells."*
   For "c," they may have written, *"I know that heavy rain makes gullies on steep slopes."*
4. a, b, and d
5. Your justification statement may have been based on your experience with eating peanut butter: *peanut butter is sticky and hard to swallow.*

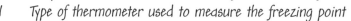

High Stakes Testing

A sample multiple-choice item from State and Standardized Exams.

A student wishes to test the hypothesis that adding antifreeze to water lowers the freezing point of the water. What would be the dependent (responding) variable?

   F   *Amount of water put into a container*
   G   *Amount of antifreeze added to the water*
   H   *Temperature at which the water/antifreeze mixture freezes* ←
   I   *Type of thermometer used to measure the freezing point*

Virginia: SOL Grade 8 Spring 2001 Relaease Item.

---

askeric.org/cgi-bin/printlessons.cgi/Virtual/Lessons/Computer_Science/EDT0013.html
www.sasked.gov.sk.ca/docs/elemsci/g4fslc12.html
www.accessexcellence.com/21st/TL/filson/formathypo.html
askeric.org/Virtual/Lessons/Science/SCI0006.html

WEBSITES FOR ACTIVITIES

A Model for Assessing Student Learning

Assessment Type: Open Response Question

Directions to the Student

Chris wants to find out which of two spot removers is better. First, he tried Spot-Remover A on T-shirts that had fruit stains and chocolate stains. Next, he tried Spot Remover B on jeans that had grass stains and rust stains. Then he compared the results.

    a. What did Chris do wrong that would make it hard for him to decide which spot remover is better?

    b. If you wanted to help Chris find out which spot remover is better, how would you design an experiment?

Open-Response [1]

_____

_____

_____

Points	Scoring Guide
4	Student recognizes kinds of material, stains, and spot removers as three variables to consider. Describes control of material and stain, while varying only spot remover. Clearly identifies a plan that will make comparison of the spot removers possible.
3	Student identifies both kinds of materials and stains as variables that must be held constant. May describe a design that eliminates variables, rather than ways of holding them constant. Misses some points of logic.
2	Students saw that Chris introduced variables in his design that were not held constant. One of those variables is identified, but a way of holding the variable constant may not be described, or it may be described incorrectly.
1	Student recognizes the problem of stain removal, but makes no comment on design. Does not recognize that variables have been introduced.
0	Blank
	Examples of Student Response* for Each Scoring Guide Level
4	a. He didn't try Spot remover A on the jeans and he didn't use spot remover B on the T-shirts. b. First I would have two T-shirts with fruit stains and chocolate stains on it. Then I would put stain remover A and stain remover B on it which ever got the stains out better would be the one he would use. I would have two pairs of jeans with grass and rust stains and I would put stain remover A on one of them and stain remover B on the other. What ever gets out the stains the best would be best.
3	a. He used two different kinds of clothing, and the clothing also had different kinds of stain on them. b. I would have the same kind of clothing and I would also have the same of stains on my clothing.
2	Chris used he spot removers on different stains. I would get spot remover a and spot remover b and use them on the same stains. The find my results.
1	I would do the same thing that he did but look a the close very carefully.

    *Student errors have **not** been corrected.

[1]Source: Grade 4, Science Question 2, KIRIS Common Open-response item. Kentucky Department of Education, KIRIS Division, 500 Metro Street, Frankfort, KY, 40601.

Chapter 13

Constructing Hypotheses

National and State Standards Connections ------------------

- Clarify questions and inquiries and direct them toward objects and phenomena that can be described, explained, or predicted by scientific investigations. (NSES 5–8)
- Formulate questions that can be addressed with data. (NCTM Pre-K–2)
- Tell about why and what would happen if? (Massachusetts Curriculum Framework, PreK–2)

Purpose

An investigation or experiment usually begins with a problem that needs solving, a question that needs answering, or a decision that needs to be made. The integrated science process skills are problem-solving and decision-making tools used to gather information (data) and test inferences (explanations). We investigate to determine if cause and effect relationships exist between things. By deliberately changing one variable in an investigation, another may change as a result. Before any investigation or experiment is conducted, a *hypothesis* is usually stated. The hypothesis provides guidance to an investigation about what data to collect. In this chapter you will learn to write hypotheses and you will use this skill later when you plan and carry out your own investigations.

Objectives

After studying this chapter you should be able to construct a hypothesis when provided with a problem.

APPROXIMATE TIME FOR COMPLETION
25

Activity 13.1 Identifying Variables to Test

Try to think of all the variables that might possibly affect how fast a person can run a 100 meter dash. There are factors related both to the individual and to the environment of the individual that could affect his speed. For example, lung capacity, muscle tone, length of legs, and motivation are characteristics of an individual that could affect his speed. The direction of the wind, surface of the track and type of shoes worn are characteristics of an individual's environment which also could affect his speed. If any of these factors could be changed, the outcome (the speed of the runner) might be affected.

Answers can be found in the Self-Check that follows.

1. Here is another problem for you to analyze:

How fast will an object fall through a liquid?

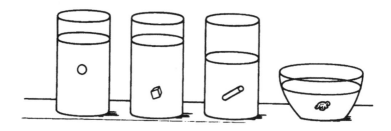

Identify variables that might affect how fast an object falls. First, consider variables related to the object, then consider variables related to the environment of the object; in this case, the liquid and the container. If materials are available in your supply area, try dropping different objects into different liquids.

What could you change about the object that might affect its speed as it falls through the liquid?

_____   _____

_____   _____

2. *What could you change about the liquid and the container that might affect the object's speed as it falls through the liquid?*

Liquid	**Container**
_____	_____
_____	_____
_____	_____

3. Another problem for you to analyze is given below. Identify some variables that might affect the growth of the plants. Remember to consider characteristics of both the plant and its environment.

What affects the growth of plants?

List some variables that might affect the plant's growth.

_____   _____

_____   _____

_____   _____

4. Identify some variables that might affect exercise time.

    What determines the amount of time an animal will spend in an exercise wheel?

List some variables that might affect the exercise time.

Animal	**Food**	**Cage**
_____	_____	_____
_____	_____	_____
_____	_____	_____

Self-Check _____ Activity 13.1

1. **Object**
    shape
    density
    volume
    mass
    composition

    These are just a few. There are many others.

2. | **Liquid** | **Container** |
    |---|---|
    | temperature | height |
    | amount | shape of container |
    | depth | diameter |

    Only some of the possibilities are listed. You may have thought of others.

3. | **Plant** | **Light** | **Water** |
    |---|---|---|
    | type | amount | amount |
    | age | direction | type of dissolved minerals |
    | quantity | color | temperature |
    | | duration | frequency |

    You may have thought of variables related to soil, fertilizer, the container, and others.

4. **Animal**	**Food**	**Cage**
age	amount	area
sex	feeding time	shape
mass	type	material

✔

 Activity 13.2 Stating a Hypothesis

Once a variable has been selected, a testable hypothesis can be stated. The term *testable hypothesis* is used because this indicates one of the functions a hypothesis should serve. A hypothesis should point the way towards the design of an experiment to test it. To construct a hypothesis, express what you think will be the effect of the variable you will deliberately change on the variable you expect to respond. This prediction can be based on knowledge, previous experience, or even a hunch. It should not, however, be a wild guess. Although some teachers use the phrase *educated guess* to describe a hypothesis, you might find that many of your young students can not distinguish between *educated* and *wild* guesses. For this reason, you may wish to ask students to add to their hypotheses a short statement that provides some rationale for their particular hypothesis. This statement will provide you some valuable insight into their thinking—both their understanding and their misconceptions. For example, to construct a hypothesis related to the problem, *What affects the speed of a car?,* a student might select the variable *diameter of tires* to test. He could then construct a hypothesis such as, *As the diameter of a car's tires increases, the maximum speed of the car decreases.* His reason might be that big tires have more contact with the road and thus would create more friction that would slow the car.

Recall in Chapter 12 that a structured format was suggested for helping students write a hypothesis:

If the _____   _____ *,*
            (independent variable)   (describe how it will be changed)

then the _____ will _____.
            (dependent variable)   (describe the effect)

Another format for writing a hypothesis is:

As the _____ *increases/decreases,*
                    (independent variable)

the _____ *increases/decreases.*
            (dependent variable)

These two formats will work for most situations, but are just two of several ways to write a hypothesis.

In the following problems, construct a hypothesis for each variable selected for testing.

Problem: Russell raises bees and noticed that different numbers of young hatched from the same number of hives at different times. He wondered what factors might influence the hatching rate of bees. He selected the following variables to test.

a. temperature of the hive
b. relative humidity inside the hive
c. amount of food available
d. number of bees living in the hive

Answers can be found in the Self-Check that follows.

1. Construct a hypothesis for any 2 of the 4 variables listed above. Add a statement that explains your reasoning.

• _____

_____

• _____

_____

Another problem:

> ***What factors determine the rate at which
> an object falls through air?***

Possible variables to test:

a. volume of object
b. surface area of object
c. length of fall
d. mass of object

2. Construct a hypothesis for two of the variables and add a statement that explains your reasoning.

• _____

_____

• _____

_____

Self-Check

1. a. As the temperature of the hive increases, the number of bees that hatch increases. (Warmth speeds up activity.)
    b. As the relative humidity inside the hive increases, the number of bees that hatch decreases. (Moisture slows activity.)
    c. As the amount of food available decreases, the number of bees that hatch increases. (To increase the number of food gatherers.)
    d. As the number of bees living in the hive increases, the number of bees that hatch decreases. (To prevent crowding.)

    These are just a few of many hypotheses that could be made. The reasons given by students as explanations for their hypotheses may be incorrect but will always be useful to you in helping them improve their understanding and correct their misconceptions.

2. a. If the volume of an object increases, then the rate at which it falls through the air decreases. (There is more friction with the air.)
    b. As the surface area of an object increases, the rate at which it falls through the air decreases. (There is more friction with the air.)
    c. The longer or farther an object falls through air, the faster it will fall. (It picks up speed as it falls.)
    d. The more mass an object has, the faster it will fall through air. (Gravity pulls on heavy objects more.)

    Your hypotheses may be entirely different from these and still be correct. In each case, however, your hypothesis should include the predicted effect of one variable upon another variable. When tested, not all hypotheses will be supported by the data. In addition, the reasoning that led to a particular hypothesis may be in error, such as in letter d.

 Activity 13.3 Putting It All Together

Now, use the following problems to practice putting together the skills you have learned:

- Selecting variables
- Constructing hypotheses
- Stating a reason for your thinking

Problem:

Why is it warmer in one house than another?

Answers can be found in the Self-Check that follows.

1. Select any two variables that might affect the warmth of houses. Write a hypothesis for each and state a reason for your particular hypotheses.

   Variable 1: _____

   Hypothesis: _____

   Reason: _____

   Variable 2: _____

   Hypothesis: _____

   Reason: _____

Here is another problem.

What factors determine the length of a shadow?

2. Generate a list of possible variables, choose two and construct a hypothesis for each related to the length of a shadow. Add a statement that explains your reasoning.

   Variable 1: _____

   Hypothesis: _____

   Reason: _____

   Variable 2: _____

   Hypothesis: _____

   Reason: _____

Self-Check _____ Activity 13.3 ✔

1. *Slope of roof:* The steeper the roof, the higher the temperature inside the house. (Larger angles capture more direct sun rays.)

   Thickness of insulation: If the insulation is thicker, then the temperature inside the house will be higher. (Insulation keeps heat in.)

   Number of windows and doors: As the number of windows and doors increases, the lower the temperature will be inside the house. (Windows and doors provide air leaks.)

   Location of the house: The nearer the house is to the equator, the higher the temperature inside the house. (Warm outside air leaks in.)

These are only a few of the many possible hypotheses that could be constructed about the warmth of houses. Your hypotheses may be very different from these. However, each hypothesis must state how you think the independent variable will affect the dependent variable. You should also include at least a little information about why you hypothesized as you did. Good teachers model the behaviors they want their students to have.

2. *Height of object:* If the height of an object increases, then the length of its shadow also increases. (Taller objects will block more sunlight.)

   Time of day: As the time moves toward noon, the shorter the shadow of an object. (Because the sun is closer to being overhead.)

   Season of year: As the seasons progress from summer to winter, the length of a shadow becomes longer. (The sun's rays strike Earth at different angles during different seasons.)

✔

You have now had an opportunity to construct several hypotheses. A second task required to design an experiment is deciding how you will measure the variables you have selected. This measurement problem will be discussed in the next chapter. Finally, in the last two chapters you will put it all together and use skills learned in all these chapters to design and conduct your own experiments. Now take the Self-Assessment for Chapter 13 before moving on to the new chapter.

Self-Assessment   Constructing Hypothesis

For each of the following problems list two variables that could affect the dependent variable. State a hypothesis for each variable listed and a statement that explains your reasoning.

1. What affects the rate of an animal's breathing?

    Variable 1: _____

    Hypothesis: _____

    Reason: _____

    Variable 2: _____

    Hypothesis: _____

    Reason: _____

2. What affects how high a balloon will rise?

    Variable 1: _____

    Hypothesis: _____

    Reason: _____

    Variable 2: _____

    Hypothesis: _____

    Reason: _____

Self-Assessment Answers

Your hypotheses may be entirely different from these and still be correct. In each case, however, your hypothesis should include the predicted effect of one variable upon another variable. When tested, not all hypotheses will be supported by the data. In addition, the reasoning that led to a particular hypothesis may be in error.

1. *amount of exercise*

   As the amount of exercise increases, the breathing rate will increase. (Because the body needs more oxygen.)

   age of the animal

   As the age of the animal increases, the breathing rate will decrease. (Because its body processes are slowing down.)

   temperature of environment

   As the temperature of the environment decreases, the breathing rate will increase. (The body has to work harder to stay warm.)

   body size

   As an animal's body size increases, the breathing rate increases. (To provide more oxygen.)

   altitude

   If the altitude increases, then the animal's breathing rate increases. (There is less oxygen at higher altitudes.)

2. *size of balloon*

   The larger a balloon is, the higher it will rise. (Because there is more surface area for the outside air to push on.)

   weight of balloon

   The lighter a balloon is, the higher it will rise. (Heavier air below pushes the balloon up.)

   temperature of air

   The cooler the air surrounding the balloon, the higher it will rise. (Cold air is heavier than warm air.)

   temperature of balloon

   The warmer the balloon, the higher it will rise. (Because warm air rises.)

High Stakes Testing

A sample multiple-choice item from State Standardized Exams.

Use the drawing below to answer the question.

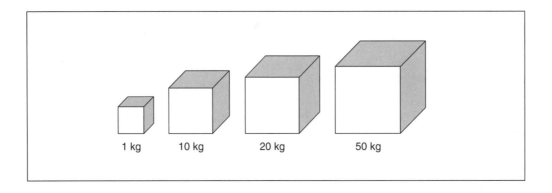

1 kg   10 kg   20 kg   50 kg

All of the boxes shown above can slide easily. Which box will move farthest if each is struck with the same amount of force?

A.   *the 1 kg box* ←
B.   *the 10 kg box*
C.   *the 20 kg box*
D.   *the 50 kg box*

Kentucky Core Content Test Grade 4, Release Items Spring 1999.

www.accessexcellence.com/21st/TL/filson/formathypo.html
www.accessexcellence.org/21st/TL/filson/writhypo.html
www.soc.iastate.edu/sapp/soc511.hyc.html
www.newfoundations.com/Hypothesizing/Hypothesize.html

WEBSITES FOR ACTIVITIES

A Model for Assessing Student Learning

Assessment Type: Interview of Individual Performance

Preparation: When time and personnel resources allow, an individual interview can be a powerful assessment tool. The following example of an interview assessment was called an Individual Competency Measure and is from Formulating Hypotheses, a Science—A Process Approach module. Listed below are the objectives measured by this assessment task and the materials needed.

Objectives: At the end of this module the student should be able to

1. DISTINGUISH between statements that are hypotheses and statements that are not.

2. DISTINGUISH between observations which support a stated hypothesis and those which do not.

3. CONSTRUCT a hypothesis from a set of observations.

4. CONSTRUCT a prediction based on a hypothesis.

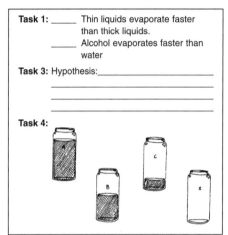

Materials

✔ Hypotheses and Predictions (to the right)

✔ Glass jars, 3, identical

✔ Water

✔ Steel rod

✔ Metric ruler

Directions

TASK 1 (Objective 1): Give the child a copy of the figure above.

Say to the students: "I will read two statements on your sheet. Put an X before the statement that is a hypothesis." Read the following statements.

1. Thin liquids evaporate faster than thick liquids.

2. Alcohol evaporates faster than water.

The child should mark the first statement.

TASK 2 (Objective 2): Say, "Assume that the hypothesis is that thin liquids evaporate faster than thick liquids. A scientist observed that water is much thinner than corn syrup and that water evaporates faster. Does this observation support the hypothesis?" **The child should reply that it does.**

TASK 3 (Objective 3): Put on a table in front of a child two identical glass jars, one of which contains a little water and the other of which is almost filled with water. Give the child a steel rod and say, "Tap both jars slightly on the sides and tell me about the sounds that are made. Be very specific when you make your observations." The child should say that the sounds are different and how they are different. If he does not say that the jar with more water has a lower pitch, ask, "Which jar has the lower pitch?" Put a third jar about half filled with water in front of the child. Say, "Tap this jar and tell me about the sound you hear. Then, order the jars from the tones of highest pitch to lowest pitch." When he has ordered the jars say, "Write on the data sheet a hypothesis about the water, the jars, and the sounds that are made when the jars are tapped with a steel rod." **The child should write something like this: When a glass jar is tapped with a rod a tone is heard. The greater the amount of liquid there is in the jar, the lower the pitch of the tone.**

TASK 4 (Objective 4): Identify the ordered jars by marking them with an A for lowest pitch, B for middle pitch, and C for the highest pitch. Say to the child, "Look at the drawings on the data sheet. Suppose I gave you a fourth jar X, that has a higher pitch than A but lower than B. Draw a line to represent the amount of water that would be in X." **The child should draw a line representing an amount of water in jar X that is between the amounts shown in jars A and B.**

[1]Module 70, Formulating Hypotheses, Science—A Process Approach II. Copyright © 1986 by Delta Education, Hudson, NH. Used by permission.

Defining Variables Operationally

Purpose

During an investigation measurements of the variables are made. Before making the measurements, however, the investigator must decide how to measure each variable. In this chapter you will practice making decisions about how to measure variables.

Objectives

When you finish this chapter you should be able to:

1. State how the variables are operationally defined in an investigation when given a description of the investigation.
2. Construct operational definitions for variables.

APPROXIMATE TIME FOR COMPLETION 25

Operationally defining a variable means describing specifically how that variable will be measured. An operational definition describes what is observed and how it is measured.

Many researchers in sports medicine and exercise physiology study the effects of various variables on the *endurance of a person.* However, each researcher might decide on a different way to measure this variable. For example, suppose an investigation was conducted to test the effects of Vitamin E on the endurance of a person. The dependent variable, endurance of a person, might be operationally defined in any of the following ways:

- The distance a person could run without stopping.
- The number of hours a person could stay awake.
- The number of jumping jacks a person could do before tiring.

Researchers choose an operational definition of the variable that will yield the best data in determining the effect of Vitamin E.

Activity 14.1 Deciding How Variables Can Be Measured

Read the following description of an investigation and determine how each of the variables was operationally defined in this investigation. That is, how were the independent and dependent variables measured?

Answers can be found in the Self-Check that follows.

1. A study was conducted to determine if safety advertising had any effect on automobile accidents. Different numbers of billboards were put up in Cleveland over a period of four months to see if the number of people hospitalized because of auto accidents was affected. In March, five billboards carried safety messages; in April there were ten; in May there were fifteen; and in June there were twenty. During each of these four months, a record was made of the number of people hospitalized because of accidents.

The independent variable, ***amount of safety advertising,*** was manipulated in this investigation to see if the ***number of automobile accidents*** would respond. How was each one operationally defined?

Safety advertising _____

_____

_____

Automobile accidents _____

_____

_____

2. A study was conducted to determine the effect that the amount of exercise has on heart rate. High school students rode bikes for different numbers of kilometers and then their heart rate was measured. One group rode 10 km, a second group rode 20 km, a third group rode 30 km, and a fourth group rode 40 km. Following the exercise the heart rate was immediately measured by counting the pulse for 30 seconds at the wrist.

How was each variable operationally defined in this investigation?
Independent variable: _____

_____

Dependent variable: _____

_____

3. A study was conducted to see if the amount of erosion was affected by the slope of the land. The end of a stream table was raised to four different heights (10 cm, 20 cm, 30 cm, 40 cm) in order to make it slope different amounts. (A stream table is a plastic box about 40 cm wide and 100 cm long. Sand or soil is placed in the box and water can be run in at one end.) At each height a liter of water was poured in at one end of the stream table. After the water had run over the soil, the depth of the gully cut by the water was measured.

Independent variable: _____

_____

Dependent variable: _____

_____

4. To think of a variety of ways that a variable might be operationally defined, consider this case. Suppose you wanted to operationally define the variable *size of a person*. Write at least three ways this variable could be defined operationally.

   1. _____

   2. _____

   3. _____

5. Now try another variable. Suppose you are an agricultural expert and are growing beans in an experiment. You need to operationally define the variable *amount of plant growth*. Write three different ways that you could operationally define this variable. Try thinking of different ways you could measure how much the plant grew.

   1. _____

   2. _____

   3. _____

6. Suppose that an elementary school has a special program to increase students' *enjoyment of reading*. What are some of the ways that enjoyment of reading could be operationally defined? List at least three ways by thinking of some specific student behaviors you could measure in your classroom that might indicate their enjoyment of reading.

   1. _____

   2. _____

   3. _____

Self-Check _____ Activity 14.1 ✔

1. The amount of safety advertising is operationally defined as the number of safety billboards put up in the city during each month. (Observed = safety billboards. Operation performed to measure what is observed = counting the number of billboards erected each month.)

   The number of automobile accidents is operationally defined as the number of people who are hospitalized because of automobile accidents. (Observed = people who are hospitalized because of auto accidents. Operation performed to measure what is observed = counting the number of hospitalized people.)

   It is important to note that these variables, or any others, could be measured in a variety of ways. It is entirely up to the investigator how the variables in the study will be operationally defined. However, the operational definition should be explicit enough that another person could replicate the measurement without any further information from the investigator.

2. *Amount of exercise* is the independent variable. It was measured by counting the number of kilometers a person rode. *Heart rate* is the dependent variable. It was measured immediately following exercise by counting the number of pulses felt at the wrist in a 30-second period. Note that an operational definition of a variable very clearly states how that variable will be measured.

   Amount of exercise and heart rate could have been operationally defined in other ways. For example, amount of exercise could have been defined by having the students run in place for designated periods of time. It could also have been operationally defined in terms of the number of knee bends each student did. There is a variety of ways you could define amount of exercise. When you select a clearly defined way to measure amount of exercise for your experiment, you have defined it operationally.

3. *Amount of erosion* is the dependent variable. It was operationally defined in this study as the depth of the gully cut by the water. Slope of the land, the independent variable, was operationally defined as the height to which the end of the stream table was raised.

   These two variables could have been operationally defined in other ways as well. For example, slope of the land could be measured in degrees of tilt of the box. Measuring the mass of the eroded soil would be another way of operationally defining amount of erosion.

4. Here are various operational definitions for *size of a person.*
   - the reading in kilograms on a bathroom scale
   - the height of a person measured from the floor to the top of his head
   - the amount of water that overflows when a person is submerged in a full bathtub
   - the amount of tape required to encircle the chest, waist and hips

5. Some possibilities you might have thought of are:
   - Count the number of leaves on a plant every week for a month.
   - Measure the distance from the soil to the uppermost leaf.
   - Find the mass of the plant and its pot. Wait one month and repeat. The difference is how much it grew.

6. You might measure the variable *enjoyment of reading* in some of these ways:
   - amount of time students voluntarily spend at the reading table
   - number of references to books read during sharing time
   - number of voluntary book reports
   - number of books taken home

Some of these operational definitions may be better measures than others. If you were doing the experiment, you would select the one you thought best.

✔

You probably realize by now that most operational definitions are determined by the person who will use them. When you conduct an investigation, you have to decide how to measure the variables. You will be constructing operational definitions when you make these decisions.

Activity 14.2   More Practice Writing Operational Definitions

In the last part of this chapter you will be given three more variables to operationally define. For each you should try to think of a variety of ways to operationally define the variable. The answers for this activity can be found in the Self-Check that follows.

1.   Concern for the Environment

Suppose that one of the goals of Springhill Elementary School is that all children acquire a concern for their environment. What are some of the ways they might operationally define this variable? Describe at least three.

1. _____

2. _____

3. _____

2.   Understanding of Fractions

Suppose that you are a fifth grade teacher and you want your students to *understand fractions*. What are some of the ways that you might operationally define this variable? Describe at least three.

1. _____

2. _____

3. _____

3.   Amount of Evaporation

An investigation is underway to see how the initial temperature of a liquid affects the *amount of evaporation*. Describe at least three ways that *amount of evaporation* could be operationally defined.

1. _____

2. _____

3. _____

Self-Check _____ Activity 14.2

1. Some possibilities are:
    - the number of special projects students choose to do on environment matters
    - the pounds of trash picked up on the playground each week
    - the number of brown bags thrown away (instead of reused) from the lunch room
    - the number of paper towels used in the washrooms
    - the number of posters on environmental matters entered in a *show-your-concern-with-a-poster contest*

Obviously there would be many ways the variable *concern for environment* could be operationally defined. Very likely you had different and possibly better ideas than these.

2. Some possibilities are:

In each case it is assumed that the students will be given the opportunity to demonstrate **understanding.**
- Multiply fractions.
- Represent fractions with cardboard pie pieces.
- Select numbers that are fractions from a list.
- Reduce fractions to lowest terms.
- Solve real-life problems involving fractions.

These are just a few ways that you might operationally define *understanding fractions* for a fifth grader. Some of them may be more valid than others. The ability to multiply fractions, for example, may be a measure of understanding of fractions, but it may also just represent memorization of a solution procedure, such as "invert and multiply."

3. Three ways that you might have thought of are:
- Measure the depth of the liquid after twenty-four hours.
- Pour a known quantity of liquid into an open container and measure its volume three hours later.
- Determine the difference between the initial mass of the liquid and the mass after 12 hours.

✔

The following is a summary of what you should have learned in this chapter.

An *operational definition* describes how to measure a variable. It should state what operation will be performed and what observation will be made. There are usually several ways that one might choose to operationally define a variable. The definition you select depends on your expectations in an investigation.

Now take the Self-Assessment for Chapter 14 before turning to Chapter 15.

Self-Assessment   Defining Variables Operationally

1. Which of the following could be operational definitions for the variable *knowledge of trees?*
   - ❑ a. identify at least fifteen different trees on a nature hike
   - ❑ b. measure the average size of trees
   - ❑ c. list at least twenty different trees that are native to your state
   - ❑ d. match picture of trees with names on a test
2. How are the variables *amount of a liquid* and *solubility of salt* operationally defined in this investigation?

An investigation is performed to see if the *amount of liquid* has any effect on the *solubility of salt* in it. (Solubility refers to the capability of dissolving a substance.) Four different amounts of water (50 mL, 100 mL, 150 mL, 200 mL) are placed in identical containers. Salt is added, five grams at a time, to each container. Each is stirred until no salt crystals can be observed in the liquid.

Amount of liquid is _____

_____

Solubility of salt is _____

_____

3. Describe three ways that you could operationally define the variable _size of automobile._

   1. _____

   2. _____

   3. _____

Self-Assessment Answers

1. a, c and d could each be an operational definition.
2. _Amount of liquid_ is measured in milliliters of water used.
   Solubility of salt is measured by mass of the salt dissolved.
3. Three possibilities are:
   a. Count the number of seats in the car.
   b. Measure the distance in meters between the front and rear bumpers.
   c. Measure the distance in centimeters between the two front tires.

High Stakes Testing

A sample multiple-choice item from State Standardized Exams.

If a scientist wanted to find out how tall a plant grows each day, the scientist would

   A.   _give the plant a half-cup of water each day._

   B.   _put the plant in a sunny place each day._

   C.   _measure the plant with a ruler each day._   ←

   D.   _put the plant on a scale each day and weigh it._

Illinois: ISAT: Grade 4 Sample Test.

www.reallygoodauthorsonline.com/ScienceWorks/the_book/5b_advanced_skills/define_oper.htm
home.inreach.com/jdcard/cogs2300/opdefn.htm

WEBSITES FOR ACTIVITIES

A Model for Assessing Student Learning

Assessment Type: Open-Response Question

Preparation

One characteristic of some forms of assessments is the blurring of distinctions between instructional activities and assessment items. To use *Balloon Rockets* as an assessment task, provide students with a copy of Activity 11.2, The Balloon Rocket, found in Chapter 11 on page 212, and a copy of the following directions.

Directions to the Student

Use *The Four Question Strategy*[1], like you used in class on other topics, to brainstorm a list of variables on the question, *What affects the movement of a balloon rocket?* Questions 1 and 2 are already completed for you. You complete Questions 3 and 4.

1. What materials are readily available for conducting experiments on *balloon rockets?*

   Response: balloons, straws, string

2. How do *balloon rockets* act?

   Response: balloon rockets move

3. How can you change the set of *balloon rocket* materials to affect the action?

4. How can you measure (operationally define) the response of *balloon rockets* to the change?

Scoring

Students' responses should include at least two ways each that the materials in Question 3 could be changed or varied, and in Question 4 at least two ways rocket movement can be measured (operationally defined). Acceptable answers would include:

3. How can you change the set of balloon rocket materials to affect the action?

   Response:

Balloons	**Straws**	**String**
diameter	brand	length
length	type of material	type of angle
type of material	diameter	tightness
# of puffs	length	type of material
shape	mass	smoothness

4. How can you measure or describe the response of balloon rockets to the change?

   Response: Measure the time the rocket takes to cross the room.

   Measure the distance the rocket travels along the string.

   Time how long the rocket is moving.

   Figure out how fast the rocket travels in meters per second.

[1]Cothron, J., Giese, R., and Rezba, R. (2000). Students and Research: *Practical Strategies for Science Classrooms and Competitions.* Dubuque, Iowa: Kendall/Hunt Publishing Company.

Chapter **15**

Designing Experiments

Purpose

In this chapter you will practice designing experiments to test hypotheses. Your skill in designing an experiment will be limited only by your imagination. However, this does not mean that your design must be complicated. Quite the contrary, the simpler the design, the more likely you will be able to collect usable data.

Objectives

When you finish this chapter you should be able to design an experiment to test a given hypothesis.

An experiment can be defined as a carefully designed procedure to test a hypothesis; the procedure is planned to yield data that will either support or not support a hypothesis. If the manner in which a variable can be manipulated and the type of response expected is clearly stated in the hypothesis, then much of the work in planning how to collect data has been done. There remains the task of specifying conditions under which the work will be conducted.

Most microwave ovens have dials or buttons numbered from 1 to 10, indicating a range of power from low to high. Have you ever been curious about just what those 'powers' really mean and what they do? Is a power level '4' twice as powerful as a level '2'? Will a food item cooked at level '8' get twice as hot as one cooked at level '4'? To find out you can design and conduct an experiment.

Here is your hypothesis: *The greater the power number selected on a microwave oven, the higher the temperature a food item being cooked will be.*

The following experimental design could be used to test this hypothesis:

Fill a large pitcher with at least 1 liter of water, letting it reach room temperature. From the pitcher, fill a microwave-safe container with water, measure and record its temperature. Remove the thermometer from the container and microwave the container of water on power '2' for 1 minute. Measure and record the temperature of the water in degrees Celsius. Calculate and record the difference between the beginning and ending temperatures. Repeat this procedure for power levels 4, 6, 8, and 10. Use the same amount of room temperature water each time. Let the container return to room temperature if it becomes warm. Allow the thermometer to adjust for 15 seconds before reading. Repeat the entire experiment two more times for a total of 3 trials, and calculate the mean temperature for each power level.

Note that this design is written clearly and succinctly so another person could actually follow it. In the following activity you will use the microwave investigation to help you identify the essential components of good experimental design.

Activity 15.1 **Identifying Essential Experimental Design Components**

Use the checklist below to identify the essential components of experimental design in the description of the microwave oven investigation.

_____ 1.   an operational definition of the independent (manipulated) variable
_____ 2.   an operational definition of the dependent (responding) variable
_____ 3.   a description of what factors are kept constant
_____ 4.   the levels of the independent variable selected for the investigation
_____ 5.   a description of the procedure to be followed
_____ 6.   a description of the trials to be conducted (when repeated trials are feasible)
_____ 7.   a control (a standard of comparison)

Self-Check _____ Activity 15.1

You should have found all the components of good experimental design in the microwave oven investigation except perhaps the control. As you may recall from Chapter 11, most experiments include a standard of comparison called a *control* or *control group*. In some experiments, such as the effect of amount of fertilizer on plant growth, the control might be the group of plants that received no amount of fertilizer. This type of control is called a *no treatment* control. In other experiments all groups receive a treatment (some amount of the independent variable). The experimenter selects one of the levels of the independent variable to serve as the control group. The level selected is often the normal or typical case. For example, in an experiment on the effect of depth of seed on seed germination, the control might be the recommended or normal planting depth; other planting depths might be deeper or shallower than the control. This kind of control is called an *experimenter-selected* control. In the microwave oven investigation the researcher might select 'level 2' to serve as the control because it is the base level to which comparisons can be made with the results from the other power levels.

Of all the components of an experiment, the concept of a standard of comparison or control is the most difficult for young students. The control or control group in an experiment is quite clear in some experiments but subjective in others. Other concepts like the dependent variable and repeated trials are easier for most young students to grasp.

✔

By now you should be good at identifying independent variables, dependent variables, and constants. Before you can design your own investigations, though, you have more to learn about other design components such as using appropriate levels of the independent variable, writing good procedures, and including repeated trials.

Although most of the investigations used in this book have five or six levels for the independent (manipulated) variable, this number is by no means sacred. It is important, however, that the investigator gather enough data to test the hypothesis and establish the relationships between the variables. The investigator must decide how many different levels of the independent variables are appropriate and how they should be selected. For example, if you were interested in finding out what effect temperature has on growing plants, you would probably not select 0°, 1°, 2°, 3°, 4 °C as the temperatures in which to grow plants because all of these levels are very close to freezing. Instead you would probably want to select levels from about 10 °C (~50 °F) to 40 °C to (~120 °F) in order to measure temperature effects on plants growing in a wide range of conditions, some colder than typical and some hotter.

Good, clear communication is essential when designing investigations. The procedure you write must be clear and succinct to the extent that someone else can follow and replicate your experiment. In Chapter 2 you practiced communication skills. Now you need to put good communication skills into practice by writing clear procedures.

Activity 15.2 Writing Procedures

In this activity you will write a procedure for someone else to follow. The person following your written description is to do **exactly what you write.** They should make no assumptions about what you might have intended to happen. Select one of the actions listed below and write a procedure for it. Then have a partner follow that procedure.

- How to measure the temperature of a glass of water.
- How to sharpen a pencil.
- How to tie a shoe.
- How to put on and button a coat.
- How to make a peanut butter and jelly sandwich.

_____

_____

_____

_____

_____

Self-Check ———————————————————— Activity 15.2

How well did the person complete their assigned task without any additional clarification? What modifications would you make to your written procedure?

Activity 15.3 **Analyzing Investigations**

In this activity you will read about a very specific situation, one that presents a problem or question that a student is truly curious about:

Earlier in the week Jacob, a middle school student, had mowed the yard. He picked up the grass clippings, bagged them and then moved them to where they would be picked up for recycling. In moving the bags Jacob accidentally put his hand inside one of the bags. It was very warm inside. He touched the grass in the other bags and they felt warm too, some much warmer than others. He was curious about just how hot a bag of clippings might get. He knew some bags had more grass clippings in them than others, so he hypothesized: *The more grass clippings in a bag, the hotter the clippings will get.*

Jacob decided to test his hypothesis. Here is his design:

> The next time the yard is mowed, put 1 bucket of clippings in a bag. Put 2 buckets in a second bag, 3 in a third, 4 in a fourth, and 5 in a fifth bag. In a sixth bag put no clippings. Tie the bags closed and place them in a shady area, separate from one another. After five days, measure the temperature in degrees Celsius inside each bag by placing a thermometer in the center of the bag (wear rubber gloves) and leave it there for one minute before removing. Immediately read the thermometer and record the temperature for each amount of clippings.

Answer the following questions about Jacob's experimental design:

Questions:

a.   How is the independent variable operationally defined?

_____

b.   How is the dependent variable operationally defined?

_____

c.   What factors will be held constant?

_____

d.   What levels of the independent variable were selected for the investigation?

_____

e.   Is there a clear procedure to follow? _____

f.   Are repeated trials planned? _____

g.   If there is a control or control group for this experiment, what is it ? _____

Self-Check _____ Activity 15.3

    a. the number of buckets of grass clippings

    b. the temperature of the grass clippings in degrees Celsius

    c. the constants include: grass clippings will all be collected on the same day; same size and type of bag will be used; bags will all be tied closed and placed in the shade; bags will be separated from one another; temperature will be measured in the same way

    d. 0, 1, 2, 3, 4, 5 buckets

    e. yes, the procedure is clear

    f. only 1 trial was described, additional trials are needed

    g. yes, the bag with no grass clippings could serve as a standard of comparison or control

Activity 15.4

Here is another example of a situation that resulted in an interesting problem for a class of students. Read their hypothesis and the design they created to test it. Answer the questions that follow the investigation design.

Situation:

    A 4th grade class used the basic recipe found at the following Internet site for making a soap bubble solution. http://www.exploratorium.edu/snacks/soap_bubbles.html

> To each gallon (3.8 liters) of water add 2/3 cup (160 mL) of Dawn™ or other dishwashing liquid and 1 tablespoon (15 mL) of glycerin, available at your local pharmacy. Bubble solution works best if it is aged at least a day before use.

    The class used the solution to make colorful dome-shaped bubbles by pouring a puddle of solution into a shallow tray, dipping one end of a drinking straw into the soap solution and gently blowing air through the straw into the puddle. Students soon found they could create a bubble that would ride on the surface of the puddle and last for at least a few seconds. They also found that when the bubble burst it left a visible white ring of soap in the puddle. Children wanted to know how they could make even bigger bubbles than the ones they had already observed. With their teacher's help they made a list of the things they might change about the bubble solution, about the straw, or about the way they blew the bubbles. They decided to try adding more soap to the basic recipe. One student explained, *"I think it is the soap that makes the bubble stretchy, so the more soap you add, the 'stretchier' and the bigger the bubbles will get.*

    The class designed the following investigation to test their hypothesis:

    Pour 100 mL of the original soap recipe into each of 5 cups. Leave the first cup as is. Add 5 mL more Dawn to the second cup, 10 mL more to the third cup, 15 mL more to the fourth and 20 mL more to the fifth cup.

    Keep everything about the original recipe the same. Use the same amount of air (just one breath) when blowing into the bubble. Hold the straw the same way each time. Measure the diameter of the soap ring the same way each time by holding a metric ruler across the widest part of the soap ring and reading to the nearest millimeter. Blow 3 bubbles with each solution and calculate the mean diameter.

Questions:

    a. How is the independent variable operationally defined?

_____

    b. How is the dependent variable operationally defined?

_____

c. What factors will be held constant?

_____

d. What levels of the independent variable were selected for the investigation? _____

e. Is there a clear procedure to follow? _____

f. Are there repeated trials? _____

Self-Check _____ Activity 15.4

a. the amount of Dawn in mL that was added to the original recipe
b. the diameter of the white soap ring, measured across the widest part of the ring in millimeters
c. same original recipe, same amount of air (just one breath) to make the bubble, holding the straw
   the same way, measuring the diameter of the soap ring the same way each time
d. 0, 5, 10, 15, and 20 mL of extra soap
e. yes
f. yes, there are 3 trials planned (3 bubbles for each solution)

Activity 15.5 Designing Your Own Experiment about Image Sizes

Now that you have learned the essential components of an experimental design, try designing an investigation yourself. Read the situation below and the students' hypothesis. Design an experiment to test their hypothesis. Be sure to include how the variables are to be operationally defined, how other factors will be held constant, and what levels of the independent variable will be used. Consider whether doing several trials is feasible and what a possible control might be. All of this should be written in one or two brief paragraphs, using complete sentences.

Situation:

   Ms. Nguyen's student helpers often set up the overhead projector for her. They could never quite remember which way or how far to move the projector to make a projected image bigger or smaller.

   They hypothesized that the *farther an overhead projector is located from the screen, the smaller an image on the screen will appear.* The students reasoned that when they look out a window the further away objects are, the smaller the objects appear.

Your Design:

Self-Check _____Activity 15.5 ✔

Exchange designs with someone else. Use the following checklist to analyze each other's experiment. Check the essential components of an experiment that can be identified in the design.

_____ 1. An operational definition of the independent (manipulated) variable.
_____ 2. An operational definition of the dependent (responding) variable.
_____ 3. A description of what factors are kept constant.
_____ 4. The levels of the manipulated variable selected for the investigation.
_____ 5. A description of the procedure to be followed.
_____ 6. A description of the trials to be conducted (if repeated trials are feasible).
_____ 7. A standard of comparison (control or control group).

✔

Activity 15.6 Designing Your Own Experiment about Electromagnets

Here is another situation with a suggested question. Read the hypothesis provided and design an experiment to test the students' hypothesis. Remember, an investigation also tests the reasoning used to make the hypothesis. Be sure to include in your design all the essential components.

Lin made an electromagnet by coiling a 25 cm length of insulated copper wire 20 times around a steel nail (about 6 cm long) and connecting the ends of the wire to the opposite ends of a C-size battery. With her electromagnet, Lin picked up small metal objects, like paperclips, then quickly released them by disconnecting one of the wires from the battery. But the electromagnet could not pick up very many paperclips at a time. Lin wanted to make it stronger. Here is her hypothesis:

If the number of batteries in an electromagnet is increased, the number of paperclips it will pick up will also increase. Lin reasoned that because batteries provide power, more power from more batteries should result in more magnetism.

Your Design:

Self-Check ————————————————————— Activity 15.6

Here is one possible design: Make an electromagnet by coiling a 25 cm length of insulated copper wire 20 times around a steel nail (about 6 cm long) and connecting the ends of the wire to the opposite ends of a C-size battery. Touch the nail to a pile of 50 paperclips. Count the number of paperclips picked up by the electromagnet. Count only those paperclips still attracted to the electromagnet after 5 seconds. Do 3 more trials with 1 battery and calculate the mean number of paperclips picked up.

Add another battery to the circuit and again count the number of paperclips picked up. Repeat the procedure each time adding additional batteries until a series of 5 batteries has been tested. Each battery used should be a newly-purchased battery, the nail, the number of coils, and the manner in which the wire is connected should remain the same throughout the experiment.

 # Self-Assessment   Designing Experiments

Choose one of the follow hypotheses and design an experiment to test the hypothesis:

1. If you increase the mass suspended from a rubber band, then there will be a directly proportional increase in the length of the rubber band.

2. As the amount of water a plant receives increases, the amount of plant growth also increases.

Self-Assessment Answers

These are just two of many possible designs.

1. Suspend the rubber band from some point. Add a hook shaped from a paperclip to the other end of the rubber band. Add 1 washer to the hook and measure the length of the rubber band between the suspension point and the top of the paperclip hook. Continue adding washers to the hook one at a time until a total of 10 washers are hanging on the rubber band. After each additional washer is added, measure the stretched distance in the same way each time. Repeat the procedure 3 more times for a total of 4 repeated trials. Use a new but identical rubber band for each trial because using the same rubber band too many times will affect its performance.

2. Obtain 25 bean seedlings each planted in similar pots with the same soil. Place them on the same window ledge in indirect natural light. Give each group of 5 seedlings a particular amount of water—10, 20, 30, 40, and 50 mL. Use room temperature tap water and water twice a week. Once a week measure their height in cm from the top of the soil to the uppermost point on each plant.

High Stakes Testing

A sample multiple-choice item from State Standardized Exams.

Dana raised two kittens. After 8 months, Dana noticed that the female cat was not as big as the male cat. Because the male cat is bigger, Dana thought that the male cat must eat more food than the female cat. To test this, Dana should measure

    A.   the amount of food the female cat eats.

    B.   the amount of food the male cat eats.

    C.   the amount of food each cat eats. ⬅

    D.   the total amount of food both cats eat.

Illinois: ISAT: Grade 4 Sample Test.

www.fed.cuhk.edu.hk/~johnson/tas/investigation/designing_investigation_teacher.htm
www.fed.cuhk.edu.hk/~johnson/tas/investigation/investigation_experiment.html
http://www.isd77.k12.mn.us/resources/cf/SciProjIntro.html
http://www.ipl.org/youth/projectguide/
http://www.muohio.edu/dragonfly/tools/research.htmlx
teachervision.com/lesson-plans/lesson-29.html?mail-03-21
www.scienceteacherprogram.org/19991p/garcia99.html
www.udel.edu/msmith/pillbugs.html

WEBSITES FOR ACTIVITIES

A Model for Assessing Student Learning

Assessment Type: Rating Sheet—Individual Performance within a Group

Directions to the Student

For each question, check the box that describes your behavior in the group during this task. Ask each person in your group to also rate your group behavior.

Name: _____ **Task Title** _____ **Date:** _____

Check one box per question.

	Almost Always	Often	Some-times	Rarely
Group Participation				
1. Joined in group discussion.				
2. Did his or her fair share of the work.				
Staying on Topic				
3. Paid attention, listened.				
4. Helped group get back to topic.				
Offering Ideas				
5. Suggested helpful ideas.				
6. Offered helpful comments on the ideas of others.				
Consideration				
7. Made good comments on the ideas of others.				
8. Gave credit to others for their ideas.				
Involved Others				
9. Helped others to join in by asking questions or asking for more information or ideas.				
10. Helped group members to reach group agreement.				
Communicating				
11. Spoke clearly. Was easy to hear.				
12. Explained ideas so they could be understood.				

Adapted from materials developed by the Connecticut Common Core of Learning Assessment Project, Connecticut State Department of Education, Bureau of Evaluation and Student Assessment, Hartford.

From Rezba, Sprague, & Fiel. *Learning and Assessing Science Process Skills,* 4th Edition. © 2003 Kendall/Hunt Publishing Co. May be reproduced by individual teachers for classroom use only.

Chapter 16

Experimenting

National and State Standards Connections -------------------

- Plan and conduct a simple investigation. (NSES K–4)
- Apply and adapt a variety of appropriate strategies to solve problems. (NCTM 3–5)
- Identify and gather tools and materials needed in an investigation. (Nevada Science Standards, Grade 3)

Purpose

Experimenting is the activity that puts together all of the science process skills you have learned previously. An experiment may begin as a question. From there the steps in answering the question may include identifying variables, formulating hypotheses, identifying factors to be held constant, making operational definitions, designing an investigation, conducting repeated trials, collecting data, and interpreting data. You will be expected to do all of these in this chapter as you plan and conduct an investigation of your own.

Objective

Following the completion of this chapter, you should be able to construct a hypothesis, design, and conduct an investigation for a problem you have identified or chosen to study.

 **Classroom Scenario**

By reading her state's science standards Maria Hernandez knew her fifth grade students should be working collaboratively to conduct experiments by observing, making accurate measurements, recording data, and communicating results through tables, graphs, and written descriptions. They needed to be designing experiments, repeating investigations, and explaining findings and any inconsistencies that might appear. Whenever possible, Ms. Hernandez involved the children in hands-on science activities that she linked by content to other areas in the curriculum. By comparing standards in all the curricular areas she began to discover similarities in the skills students are expected to learn in each of the subject areas. Making inferences and drawing conclusions in science, for example, are much the same as making inferences and drawing conclusions in reading. Calculating means while conducting an experiment is just like calculating averages like mean, median, and mode in math class.

Ms. Hernandez spent much of the school year involving students in 'partial' inquiries, activities designed specifically to develop students' abilities and understanding of parts of the inquiry process. They practiced identifying variables, writing procedures, collecting and organizing data, and drawing conclusions. The students were now ready to conduct a 'full' inquiry, one in which they would put it all together. They would ask a question, write a hypothesis, and design their own experiments to test their hypothesis. They would carry out their own design, collect data, organize it and try to make sense of what they had found.

On one particular day the children were working in cooperative learning groups. Before starting the activity, Ms. Hernandez carefully checked students' understanding of their assigned roles in the group:

Ms. Hernandez:	What does the *materials manager* do?
Deana:	They get and return materials from the supply table. And they make sure we clean up when we are done.
Ms. Hernandez:	What does the *design engineer* do?
Josh:	She checks to make sure the experiment we design makes sense and that it does what it is supposed to do to test our hypothesis.
Ms. Hernandez:	What does the *data collector* do?
Kevin:	That person does all the measuring.
Ms. Hernandez:	What does the *recorder* do?
Fredericka	He makes sure everyone agrees on everything and writes it down on paper.
Ms. Hernandez:	Lamar, go up to the supply table and tell us what you see there.
Lamar:	I see marbles—some are glass, some are plastic, some are steel, and they are different sizes. There are rulers, pieces of sponge, cups, blocks of wood, sandpaper, cooking oil, Vaseline, and stacks of books.
Ms. Hernandez:	When you are in your group, remember to use your *100-centimeter voices*. What is a 100-centimeter voice?
Juanita:	When you talk, no one farther away than 100 centimeters should be able to hear you.
Ms. Hernandez:	Who are the only children who should be out of their seats and moving around?
Josh:	Only the materials managers. They can go to the supply table.
Ms. Hernandez:	How do we know who the materials managers are?
Josh:	We know who they are because they are the only ones that have a materials manager clothespin clipped on their shirt.

Following Ms. Hernandez's instructions each group made a ramp by taking a grooved ruler and elevating one end with a stack of books. They rolled a marble down the ramp that then hit a piece of sponge positioned at the bottom of the ramp, pushing it along the tabletop. Students made some initial observations and recorded such things as the height of the ramp and the distance the sponge traveled when struck by the marble. One group observed that sometimes the sponge just got pushed to one side and the marble continued on by itself.

After a few trials, Ms. Hernandez instructed each group to write an explanation for **why** the sponge moved the way it did when it was hit by the marble. She instructed them to use the words 'energy' and 'force' that they had been studying in their explanations.

Then Ms. Hernandez asked students to think about some variables, things that could be purposely changed to somehow affect how far the sponge is pushed by the rolling marble. As the students named possible variables Ms. Hernandez recorded them on the board:

Ms. Hernandez:	What could you change about the **marble** that might affect how far the sponge is pushed?
Tara:	What if we use a bigger marble?
Jacob:	Or a steel one, or a plastic one?
Miss Hernandez:	What could you change about the **ramp** that might affect how far the sponge travels?
Juanita:	What if we make the ramp longer? We could put two rulers together.
Stephanie:	What if we stack more books to make the ramp steeper?
Tara:	What if we put oil on the ramp? Or put sandpaper on it?
Miss Hernandez:	What could you change about the **object** that might affect how far it is pushed by the marble?
Jacob:	What if the sponge is wet?
Karen:	What if we used a block of wood instead of a sponge? Or what if we lay a cup on its side and the marble rolls into it and pushes it?
Juanita:	What if we use something heavy instead of a sponge?

Ms. Hernandez instructed students to think about the suggested variables and to write one question their group could actually investigate using materials from the supply table. As each group read their question to the class, Ms. Hernandez prompted them to think about what their hypothesis might be and how they might actually test it.

Ms. Hernandez:	What question did your group choose to investigate, Deana?
Deana:	*If we raise the end of the ramp, will the sponge get pushed farther when the marble hits it?* But we want to use a cup instead of the sponge so it catches the marble and gets pushed along.
Ms. Hernandez:	That's a good question to investigate. What is your hypothesis?
Lamar:	We think the higher the end of the ramp is, the farther the cup will be pushed by the marble.
Ms. Hernandez:	Lamar, can you explain why you think a taller ramp will cause the marble to push the cup further?
Lamar:	We think the marble has more energy when it sits up higher off the table.

Ms. Hernandez moved on to the next group and asked what question they would investigate.

Sara:	Our question is, *Will putting different pieces of sandpaper on the groove of the ruler affect how far the marble pushes the sponge?*
Ms. Hernandez:	What will you deliberately change in your experiment?
Lamar:	We'll use different pieces of sandpaper that range from smooth to rough.

Ms. Hernandez:	What do you think will happen?
Jon:	We think that as the sandpaper gets rougher, the marble will move slower and not push the sponge as far.
Ms. Hernandez:	How is energy involved?
Sara:	Slower moving marbles have less energy to push the sponge.

As Ms. Hernandez moved from group to group she checked their questions and hypotheses. She remained closely involved, questioning and guiding students as they designed experiments to test their hypotheses.

Here is how one group in Ms. Hernandez's class designed their investigation and eventually answered their own question.

───────────────────────── Classroom Scenario ─────────────────────────

A Sample Experiment Report

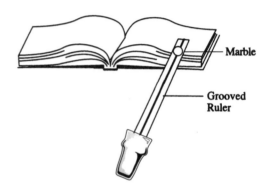
Marble
Grooved Ruler

Problem: What affects how far a marble rolling down a ramp will push an object sitting at the bottom of the ramp?

Hypotheses: The steeper the ramp, the farther a marble rolling down the ramp will push a small cup. We think the marble has more energy when the ramp is steep, so there is more energy to push the cup and it will go farther.

Design: Use a grooved ruler as a ramp for a marble to roll down. When a marble rolls down the ramp it will go into a cup and push it along the tabletop. Use books to lift one end of the ramp to different heights and measure the height of the books: 7, 9, 11, 13, 15, 17, 19, and 21 cm. Lay a meter stick on the table starting at the bottom of the ramp. Measure the distance the cup moves by reading where the lip of the cup comes on the meter stick. Each time start the marble from the same place on the ruler and release the marble in the same way. Each time place the cup tight against the bottom of the ramp so the marble will go in it and push it along. Do three trials for each measurement and calculate the mean.

Data Table:

How Does Ramp Height Affect the Distance a Cup Is Moved?

Height of the ramp (cm)	Distance the cup moved (cm)			Mean distance moved (cm)
	trial 1	trial 2	trial 3	
7	17	14	17	16
9	19	17	16	17
11	19	17	19	18
13	23	23	43	30
15	30	35	42	36
17	42	49	43	45
19	48	48	40	45
21	20	38	32	30

Graph:

How Does Ramp Height Affect the Distance a Cup Is Moved?

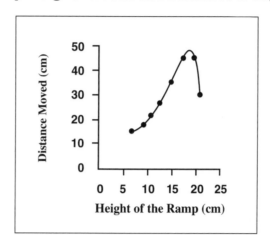

Statement of Relationship: As the height of the top of the ramp increased from 7 to 10 cm, the distance the marble pushed the cup increased gradually. From 10 to 17 cm, the distance increased rapidly, then stayed the same from 17 to 19 cm. After 19 cm the distance the cup moved decreased rapidly.

Findings: What we thought would happen, did not really happen. Our hypothesis was only partially supported by our data. We thought the cup would continue to move farther the higher we raised the ramp. But at first, raising the height of the ramp made little difference. We think that happened because there was just very little force present to move the cup. And after 19 cm the distance the cup moved actually decreased. We think that happened because the marble no longer hit the bottom of the cup directly. It was deflected off the side of the cup.

Conclusion: How far an object is moved by a force depends not only on the amount of force present but on the direction from which the force comes.

How Do You Know Whether to Use a Line Graph or a Bar Graph to Report Your Data?

In the student investigation described above a line graph was used to report the data. But how do you really know what type of graph to use?

It should be noted that in some investigations the levels of the independent (manipulated) variable are distinct types or discrete categories, such as brands of paper towels, types of wood, gender, days of the week, and kinds of batteries. Discrete means that the categories are separate and not continuous. The spaces or intervals between categories are not equal and thus, have no meaning; there is no brand of paper towel that is halfway between Brand X and Brand Y. When the independent variable consists of discrete categories, such as brands of paper towels—Brand X, Brand Y, and Brand Z—you must display the data as a **bar graph,** not a line graph.

A **line graph** is appropriate whenever the variables are continuous. Continuous means that the values of a variable are not separate categories and that the intervals between them have meaning. What affects the absorbency of paper towels? Suppose you use different amounts of paper towels, 1, 2, 3, 4 and 5 towels to see how number of towels affects the mass of water absorbed. Because number of towels is a continuous variable, you could also try 2 1/2 sheets. The interval between 2 and 3 sheets has meaning. Other examples of continuous variables are volume of water, heights of ladders, units of clock time, and mass of fruit produced.

Does the Brand of Towel Affect How Much Water Is Absorbed?

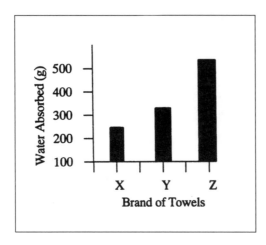

Does the Number of Towels Affect How Much Water Is Absorbed?

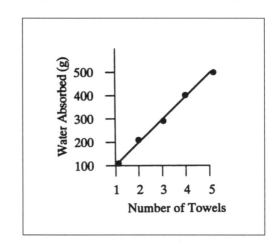

Drawing Conclusions

Like scientists, students should report their conclusions in writing. They can use words, measurements, and pictures. Some rules apply:

Words are always written in complete sentences.

Measurements are written with units.

Pictures are carefully drawn.

Re-read the students' conclusion statement in the previous investigation. Note that students did more than just report what happened in their experiment. They also tried to make sense of what they found. By having students explain **why** their results occurred, we encourage them to integrate what they already know with new information gained in the experiment. An effective teacher invites students to reflect and to speculate as to **why** things happen by asking thought provoking questions such as these:

- What was the purpose of your experiment?
- Was your hypothesis supported by the data?
- Why do you think your results happened as they did?
- What were the major findings? Include a clear statement that describes the relationship between the independent (manipulated) variable and dependent (responding) variable.
- What did you find that surprised you?
- What other factors might be affecting your results?
- How did your findings compare with someone else's or with information in the textbook?
- Why do you think your conclusion might be different from someone else's?
- What possible explanation can you offer for the findings?
- What recommendations do you have for further study?
- What recommendation do you have for improving the experiment?

Activity 16.1 **Conducting Your Own Experiment**

Now its time for you to use what you have learned in the earlier chapters. You will need a problem to investigate. You may already have a problem in mind or perhaps one has been assigned to you by your instructor. Perhaps you can locate or recall an activity that offers opportunities for manipulating variables. Such an activity can serve as a springboard into your inquiry. Here are some additional suggestions:

What affects the amount of gas produced when vinegar and baking soda are mixed together?

What affects how far a rubber band will fly?

What affects the rate at which an object falls through a liquid?

What affects the absorbency of a paper towel?

What affects the strength of an electromagnet?

What affects a person's reaction time?

What affects the amount of static electricity on a balloon?

What affects the rate at which water freezes?

What affects the length of time a person spends brushing their teeth?

What affects how fast mold grows on bread?

What affects how fast a nail rusts?

What affects how hot it gets inside a parked car?

What affects the dissolving rate of a vitamin?

How does practice (number of trials) affect the length of time it takes to do simple tasks, such as looking up a word in the dictionary, finding a location on a map, or "dialing" a phone number?

What affects the rate at which a pipe transports water from one place to another?

What affects the amount of time it takes a seed to sprout?

What affects the rate at which a person breathes?

When you have selected a problem to study, state your problem as a question. Construct a hypothesis, design an experiment to test your hypothesis, collect the data, and write a brief report. A page or two should be sufficient to contain all the information you need in your report. The report should include:

1. The statement of the problem or question you are investigating.
2. The statement of the hypothesis you are testing (Add a statement that explains your reasoning.).
3. A written description of the design of the experiment you used to test the hypothesis (Remember to describe how the variables were operationally defined, factors that were held constant, the levels of the independent variable you used, the number of repeated trials, and a description of your control.).
4. A data table that includes repeated trials and a derived quantity such as the mean.
5. A graph of the data.
6. A statement of the relationship observed between the variables.
7. A comparison of your findings with your initial hypothesis stating whether the hypothesis was or was not supported by the data.
8. A conclusion that attempts to explain what happened.

Self-Check

Use the Science Experiment Rubric on page 274 . Depending upon the type of experiment you did, you may find it desirable or even necessary to modify the criteria or possible number of points.

When you finish this investigation you are well prepared to begin teaching the science process skills to children and to involve them in conducting inquiries of their own. You may use the investigation you just completed as a model to use with your students. It is important for children to see you as a model investigator.

Hang On! More planning help is at your fingertips.

Turn to the **Appendix** for these helpful materials:

- A sample format for planning an inquiry-based lesson (Plan with confidence.)
- A sample inquiry-based lesson plan (Get started with a simple example.)
- A sample format for a student experiment report (Get students organized and thinking.)
- Sample activities and variables to study (Get ideas flowing.)

High Stakes Testing

A sample multiple-choice item from State Standardized Exams.

A bottle of water will break if it is left in the freezer overnight. Which is the fairest way to show that it is the freezing water that breaks the bottle and not just the bottle getting too cold?

F

H ⬅

G

J

Virginia: SOL 2000 Release Item Reporting Category: Scientific Investigation.

http://www.HunkinsExperiments.com/
http://www.wff.nasa.gov/~sspp/sem/sem.html
http://www.slb.com/seed/en/lab/index.htm
http://www.tryscience.org/
http://www.mcrel.org/resources/whelmers/index.asp?
http://www.muohio.edu/Dragonfly/
www.isd77.kl2.mn.us/resources/cf/SciProjInter.html
www.fed.cuhk.edu.hk/~johnson/tas/investigation/banana_student.htm
unr.edu/homepage/crowther/ejse/papertowel.html
www.popweaver.com/scifair.htm

WEBSITES FOR ACTIVITIES

A Model for Assessing Student Learning

Assessment Type: Teacher Rating Sheet

Science Experiment Rubric

Experimenter's Name: _____

Peer Reviewer's Name: _____

Criteria (*number of points given in parentheses*)	Possible Points	Peer Review	Instructor's Review
Problem: The problem to be investigated is stated as a question (1), is clear and understandable to a new reader (1), can be investigated using safe and available materials (1), will generate new learning for the investigator (1).	4 points		
Hypothesis: The hypothesis is properly stated as a prediction of what might happen (1), includes a brief explanation for WHY the prediction is made the way it is (2), describes the expected relationship between the variables involved (1).	4 points		
Design: The design includes the operational definition of the independent variable (2), sets up at least 5 different levels of the independent variable (1), includes an operational definition of the dependent variable (2), includes all the necessary constants (2), includes a clear and thorough description of the procedure to be followed (3). The procedure is designed to gather sufficient amount of data to support or not support the hypothesis (1). Repeated trials are described. (1)	12 points		
Table: The table uses proper labels for the independent and dependent variables (2 + 2). Metric units are used (1). Levels for the independent variable are sequentially ordered (1). Repeated trials are recorded (2). Mean, median, or mode is appropriately calculated (2).	10 points		
Graph: The graph has an appropriate and descriptive title (2). Both axes are appropriately labeled (2 + 2). Numbers on the axes are assigned beginning at the corner of the graph (1). Equal intervals are used along each axis (2). In a line graph the line is drawn as a best-fit line (1), the line is either a straight line or smooth curve (1), the line accounts for all data collected (1). Or for a bar graph, the bars are accurately drawn (3).	12 points		
Statement of Relationship: The statement is read directly from the graph (1), accurately describes the relationship between the independent and dependent variables (2). The statement describes all changes in relationships and tells when they occur (curves require more than one sentence to describe them (1).	4 points		
Conclusion: The conclusion attempts to explain WHY the results happened as they did (2), compares findings to the hypothesis by stating whether data supported or did not support the hypothesis (1), states limitations and possible future investigation (1).	4 points		
Total	**50 points**		

Decision Making 2

Now that you have learned the Integrated Science Process Skills, you can use what you have learned to improve existing science curricula. In learning the science process skills, you not only mastered the skills, but you also learned something about how these skills can be taught. By using this knowledge you can begin making some important instructional decisions about teaching science, especially the science process skills. In this section you will focus on the *application* of what you know about the integrated science process skills to improve elementary and middle school science textbook activities. The decisions you make can significantly enhance the quality of science in which your students are engaged.

Read *Typical Textbook Activity Example C* on the next page. Think about how you might change the activity to better emphasize the science process skills.

As you study the sample activity, look at both the content and skills your students will be learning and how they would be learning them.

Ask yourself, *How will I provide opportunities for my students to:*

- formulate or choose questions to study?
- state hypotheses?
- identify and define variables operationally?
- design investigations?
- conduct investigations and acquire data?
- construct tables of data?
- construct graphs?
- describe relationships observed between variables?
- describe the finding of an experiment?

With these questions in mind, write what you consider to be strengths and weaknesses of *Example C* on a separate piece of paper. Then consider how you might change the activity to improve the weak areas.

After you have studied *Example C* and thought about how you might change it, turn the page and look at our modified version of *Example C*. The changes made to *Example C* are only a few modifications that could be made to this activity to better emphasize the process skills. Your ideas for modifying this activity may have been different and even better.

Typical Textbook Activity Example C

ACTIVITY

You will need:

✔ 1 ruler with a groove running end to end
✔ 1 marble
✔ 1 small block of sponge, (about 2 cm x 5 cm x 5 cm)
✔ 1 other ruler, metric ruled

Can Work Be Measured?

Follow This Procedure

1. Place the grooved ruler on the table in front of you.
2. Place the sponge block at one end of a ruler so that the sponge touches the end of the ruler.
3. Raise the opposite end of the ruler 2 cm above the table.
4. Place the marble at the top of the ruler groove and let it roll into the sponge block.
5. Measure how far the sponge moved. Record your results in the table. The greater the distance the sponge moved, the more work the marble did on the sponge.
6. Repeat this activity raising the height of the ramp to 4 cm, then 6 cm, and 8 cm above the top of the table. Record your results.

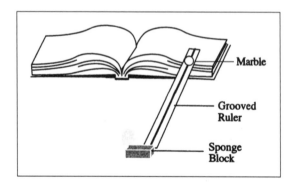

Marble

Grooved Ruler

Sponge Block

Height of Ramp	Distance the Sponge Moved
2 cm	
4 cm	
6 cm	
8 cm	

What Did You Find Out?

1. At what ramp height did the marble have the most potential (stored) energy?
2. How was the amount of work done affected by the marble's potential energy?

Here are our suggestions. There are also many other ways to emphasize the integrated science process skills in this activity, such as asking students to identify the constants in this investigation and asking them to formulate a hypothesis and then state whether the data supported or did no support their hypothesis.

Modified Textbook Activity Example C

ACTIVITY

Can Work Be Measured?

Part 1
Modify activity by adding
repeated trials.
Also add column for mean.

You will need:

✔ 1 ruler with a groove running end to end
✔ 1 marble
✔ 1 small block of sponge, (about 2 cm x 5 cm x 5 cm)
✔ 1 other ruler, metric ruled

Follow This Procedure

1. Place the grooved ruler on the table in front of you.
2. Place the sponge block at one end of a ruler so that the sponge touches the end of the ruler.
3. Raise the opposite end of the ruler 2 cm above the table.
4. Place the marble at the top of the ruler groove and let it roll into the sponge block.
5. Measure how far the sponge moved. Record your results in the table. The greater the distance the sponge moved, the more work the marble did on the sponge.
6. Repeat this activity raising the height of the ramp to 4 cm, then 6 cm, and 8 cm above the top of the table. Record your results.

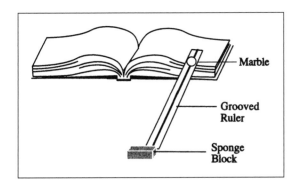

Height of Ramp *(cm)*	Distance the Sponge Moved *(cm)*			Mean distance moved
	1	*2*	*3*	
2 cm				
4 cm				
6 cm				
8 cm				

Part 2 (In groups of 4)
In post lab discussions, ask students to choose some other variable they might manipulate (size of marble, other rolling object, length of ruler runway, roughness of surface, and so on). Using their chosen independent variable, have students design and conduct own investigation. Make table of data and graph. Report results to class.

What Did You Find Out?

1. At what ramp height did the marble have the most potential (stored) energy?
2. How was the amount of work done affected by the marble's potential energy?

Here is another textbook activity example. Your task is to modify this activity to emphasize the process skills as modeled in the previous example. It may help you to review the questions on page 275 and to describe this activity's strengths and weaknesses. Then make your changes right on this activity page. When you are done, see the next page for some modifications we made.

Typical Textbook Activity Example D

ACTIVITY

You will need:

✔ a partner
✔ 1 metric ruler

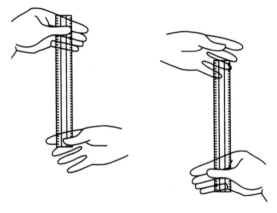

Measuring Your Reaction Time

Follow This Procedure

1. Stand or sit facing your partner.
2. Hold your index finger and thumb open while your partner suspends the end of a ruler between them. Your fingers should be at the 0 cm mark.
3. Watch the ruler closely. When your partner drops the ruler, catch it between your fingers. Record the number where your fingers caught the ruler.
4. Repeat steps 2 and 3 seven more times.
5. Graph the data you collected for all 8 trials.
6. Exchange places with your partner and repeat steps 2–5.

Write Your Conclusions

How did the number of trials affect your reaction time?

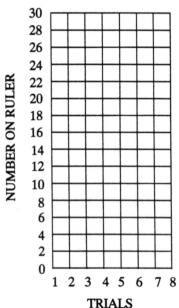

Here are some modifications to this activity that we made. Your modifications may be even better.

Modified Textbook Activity Example D

use this activity to focus on stating hypotheses and setting up the design of an investigation used to test the hypothesis. Would also use to assess students' ability to state hypotheses and design experiments.

ACTIVITY

You will need:

✔ a partner
✔ 1 metric ruler

Measuring Your Reaction Time

Follow This Procedure

1. Stand or sit facing your partner.
2. Hold your index finger and thumb open while your partner suspends the end of a ruler between them. Your fingers should be at the 0 cm mark.
3. Watch the ruler closely. When your partner drops the ruler, catch it between your fingers. Record the number where your fingers caught the ruler.
4. Repeat steps 2 and 3 seven more times.
6. 5. Graph the data you collected for all 8 trials. *Draw a line of best fit.*
7. 6. Exchange places with your partner and repeat steps 2–5.
5. *Make a table for the data collected.*

Write Your Conclusions

How did the number of trials affect your reaction time?

8. *State the hypothesis being tested by this experiment.*
9. *State another hypothesis that could be tested using this activity (comparing reaction times by gender, age, time of day, and so on).*
10. *Describe the design of an investigation that would test the hypothesis.*

NUMBER ON RULER
30 28 26 24 22 20 18 16 14 12 10 8 6 4 2 0
1 2 3 4 5 6 7 8
TRIALS

Modifying Real Textbook Activities

The textbook examples you have just studied are typical of the kinds of science activities you might find in an upper elementary or middle school textbook. To gain a little experience at modifying materials to emphasize the science process skills and to become acquainted with real textbook activities, you have one more task to complete.

Obtain either an elementary science textbook for grades 4, 5, or 6, or a middle school science textbook. Locate an activity and modify it to better emphasize the integrated science process skills just as you did in the previous examples. You might refer again to the questions on page 275 that focus on these skills. You may find it helpful to think of this task as a three-step procedure:

1. Examine the activity.
2. Identify parts that could be improved.
3. Add your improvements.

For feedback on your modifications, see your instructor, or try your modified activity with children and assess their skills.

At this point you have spent considerable time and effort in learning the science process skills. You have also been asked to think about how you will teach these skills to children and ways you might assess how well students have learned these skills. How has all of this impacted your perception of your role as an elementary or middle school teacher of science?

Turn the page and complete the section called **Growing Professionally.**

Growing Professionally and Examining Goals

Now that you have completed the activities and readings in *Learning and Assessing Science Process Skills,* return to the goal setting exercise on page xv.

Read the statements you wrote describing the achievement for which you would like to be known after teaching science for three years. If your earlier goals have changed, rewrite your statements to reflect these changes.

Finally, in the space below, write two more things you will do to continue your professional development toward becoming an effective teacher of science.

_____

_____

_____

_____

_____

_____

_____

_____

_____

_____

_____

_____

_____

_____

_____

_____

_____

_____

_____

_____

_____

Appendix

Materials and Equipment for Learning and Assessing Science Process Skills

Equipment and Materials

The materials listed below are needed to do the activities in this book. Many of the materials can be found at home or purchased locally. The most expensive item on the list is a double pan balance and masses.

Quantities are listed in amounts needed per person if students are working individually or per group if small numbers (3-4) of students are working together. No materials are listed for Chapter 16 because the materials depend on which problem is chosen to be investigated.

Internet access is desirable in order to take full advantage of the technology references listed in some chapters.

Item	Amount	Used in Chapters
plant	1	1
magnifying lens or jeweler's loupe	1	1
cornstarch	1 box	1
mixing bowl or similar container	1	1
sealable plastic sandwich bags	1 box	1, 4
35 mm film canister	1	1
paper towels		1
baking soda	1 box	1
round, transparent (deli) containers (225 or 250 mL)	1	1
plastic spoons	1 box	1,7,11
balances and masses	1	1,4
meter stick and metric ruler	1	1,4,5,6
vinegar	1 bottle	1
Gobstopper candies (or M&Ms)	1 box	1
optical or digital microscope (optional)	1	1

tray or open container to catch spills	1	1
graduated cylinder 50 mL or larger	1	1, 4, 11
sensory materials (a variety of objects to smell, feel, hear, see)	1 set	2
magnetic compass	1	2 (optional), 5
sea shells (an assortment)	1 set of six numbered 1,2,3,4,5,6	
cereal box information panels	an assortment	3
peanuts in the shell	1 bag	3
pasta shapes	an assortment	3
centicubes	20	4
liter containers	1	4,7
containers (assorted sizes and shapes)	4	4, 7
glass jar (optional)	1	6
particles to fill the jar (peas, marbles, rice, and so on)		6
board about 1 meter long and 30 cm wide to serve as a ramp	1	6
unopened and un-dented cans of food to roll down the ramp; one must contain a runny liquid (chicken broth for example) and the other must be the same height and have the same diameter as the first can and must contain chunky food (diced tomatoes or chunkysoup for example)	2	6
washers or sinkers or other small objects to serve as pendulum bobs	a variety	6
pendulum support (a pencil or ruler will do)	1	6
paper clips (to use as hooks)	1 box	6
calcium chloride (may be purchased in an automotive store, used to melt ice)	500 grams	7
safety goggles	several	7
identical containers (beakers or plastic cups)	4	7
large container (for holding at least 1 liter of water)	1	7
100 mL Pyrex beakers	4	11
measuring spoon	1	11
hot plate	1	11
timer or clock with a second hand	1	11
masking tape	1 roll	11
balloons (assorted sizes)	1 bag	11

plastic straws	1 box	11
quart size sealable bags	1 box	11
paper cup (about 3 ounces)	5	11
sharp pencil for making holes in paper cups	1	11
ice cubes	as needed	4
small, plastic disposable cup	1	4
spring scale calibrated in newtons	1	4
mesh bag (to hang from the spring scale to hold small objects)	1	4
Celsius thermometer (0–120 °)	1	4,7,11
rocks, one solid rock and one porous rock (their volumes must be large enough to displace a measurable amount of water, and their mass must be within a range that can be measured on an available spring scale)	2	4
can of soda	1	4
potato chips	1 bag	4
cookies	1 box	4
sugar	1 container	4,11
vegetable shortening that is solid at room temperature	1 container	4
overflow container (optional): students can make one using a 2 liter bottle, bendable straw, standard hole punch, scissors	1	4
small, plastic disposable cup	1	4
gravel	enough to fill a small cup	4
empty soda can	1	5
small balloon	1	5
soft cloth	1	5
mystery box	1	5
string or cord	1 ball	5, 6, 11
bar magnet	1	5

A Format for Planning an Inquiry Lesson

When students are truly involved in inquiry, they are asking questions that they can investigate. By designing and conducting their own experiments, they have opportunities to collect, record, and organize data. They learn to analyze their results by reporting findings, making sense of their data, and drawing meaningful conclusions.

Much planning goes into a successful inquiry lesson. The teacher must identify the concepts, standards, and objectives to be addressed by the lesson. Strategies are needed to assist students in asking questions that they can answer through investigation. And the teacher must identify specifically what the student must produce to show that an answer was arrived at using appropriate processes. The teacher must identify what it is that the student must produce as evidence of having answered the question. All of these decisions are made in the planning stage before the lesson begins. Other decisions, like how to design and conduct the experiment, are left to the inquiring student. The teacher can, however, anticipate how students might proceed by planning materials and some guiding questions.

While there are many ways to plan an inquiry lesson, here is a format you may find helpful to begin to organize your thinking and to provide structure as you plan an inquiry lesson.

Lesson Title: (For easy reference, give your lesson a name.)

Concepts addressed: (List the science concepts addressed in this lesson.)

Standards addressed: (Refer to state or national standards.)

Objectives: (Include objectives that describe the *processes* you want students to use in their investigation **and** include objectives that describe the *content* you want students to have learned as a result of the lesson.)

Problem/ Discrepant Event/ Question (What is the problem students will investigate, how will you present it to students, or how will you get students to ask their own question to investigate?)

Evidence of Solution, Resolution or Answer (Describe what students will produce, perhaps an experiment report, to show that they have used the *processes* you want them to use and that they have drawn a *meaningful conclusion* to their problem.)

Materials: (What materials do you anticipate students will use?)

Anticipated Lesson Development: (How do you think students will go about investigating a solution to the problem?)

Guiding Questions: (In your role as facilitator, what questions will you ask to encourage investigation, to keep students on task, and to guide their thinking toward a meaningful conclusion?)

Assessment of Student Performance: (How will you measure the quality of what students produced? Did they use the processes you wanted them to use? Did they draw a meaningful conclusion to their investigation? Did they learn the content you want them to learn?)

A Sample Inquiry Lesson

Lesson Title: Marbles and Ramps

Concept: Force and Motion, Potential and Kinetic Energy

State Standards:

Standard 1—The Nature of Science and Technology: *Students are actively engaged in exploring how the world works. They explore, observe, count, collect, measure, compare, and ask questions. They discuss observations and use tools to seek answers and solve problems. They share their findings.*

Standard 2—Scientific Thinking: *Students begin to find answers to their questions about the world by using measurement, estimation, and observation as well as working with materials. They communicate with others through numbers, words and drawings.*

Standard 3—The Physical Setting: *Students collect and organize data to identify relationships between physical objects, events, and processes. They use logical reasoning to question their own ideas as new information challenges their conceptions of the natural world.*

- Recognize and describe that energy is a property of many objects and is associated with mechanical motion.
- Recognize that objects may have energy and be capable of doing work because of their position.

Objectives

1. Given a simple science activity, the student will identify the variables involved in that activity.
2. The student will write a question about the activity that can be answered through investigation.
3. The student will design and carry out an investigation to answer their question.
4. The student will collect and organize data resulting from the investigation.
5. The student will describe relationships that occur between the independent and dependent variables.
6. The student will state the findings from his/her investigation and draw a meaningful conclusion.
7. The student will describe how energy is stored in an object because of its position, then released as the object is set in motion.
8. The student will identify at least one variable involved in how much energy an object stores and releases.

Identifying a Problem

- Have students roll a marble down a ramp (grooved ruler) where it strikes a sponge placed at the bottom of the ramp.
- Acquaint students with the materials and basic ideas related to the concepts of objects in motion and forces.
- Ask: What are some variables you could change in this activity?
- Direct students to write down at least one question that can be answered by investigation using variables just identified. Some questions students might suggest:

    What would happen if I changed the steepness (angle) of the ramp?
    What would happen if I changed the size of the marble?

What if I changed the roughness of the ramp?

What if I changed the mass of the object being pushed?

What if I used a longer ramp?

Evidence of Solution/Resolution to the Problem:

Students will submit their completed experiment report showing the processes they used to answer the question. (Distribute experiment report form and review it with students.)

Anticipated Materials:

Place on the supply table . . .

grooved rulers to serve as ramps

rulers for measuring

marbles, various diameters, mass, and composition

books (to elevate one end of the ruler)

objects (such as a sponge or cup) for a marble to hit and move as it reaches the bottom of the ramp

sandpaper of various grades

Anticipated Lesson Development:

Clarify for students how they will use the Experiment Report sheet to document their investigation. Students must:

- State the question they will investigate.
- State a hypothesis that states not only WHAT they think will happen but WHY they think it will happen as they change one variable and observe what happens to the other variable.
- Design how they will conduct their investigation making sure the variables are identified and defined operationally, constants are identified, and that the procedure communicates clearly how the investigation will be conducted.

At this point each group must have the teacher check their design before they are allowed to proceed.

- Proceed with the investigation. Collect, record and organize their data in a data table.
- Graph the data.
- Analyze their data and state the relationships found.
- Describe their findings by relating what they found to the hypothesis.
- Draw a meaningful conclusion.

Some Guiding Questions to Ask Students:

What forces are at work when the marble is at rest at the top of the ramp? What forces are at work when the marble rolls down the ramp? What forces are at work when the marble hits the object?

Where does the energy come from that is stored in the marble when it is sitting at the top of the ramp?

What happens to the stored energy as the marble rolls down the ramp and strikes the object?

Why does the object move when struck by the marble?

Students clean up their work area and return all materials to the materials table.

Assessment of Student Performance:

1. Collect students' Experiment Reports. Apply the Experiment Rating Sheet and Rubric (on page 274) to assess students' use of process skills.
2. Examine students' conclusions to assess their understanding of content. Apply the following content rubric to students' conclusion statements.

Content Rubric

Criteria *(possible number of points in parentheses)*	Points Awarded
Student explains that energy is stored in the marble because of its position at the top of the ramp. (1)	
Student explains that energy stored in the marble is released when it travels down the ramp. (1)	
Student describes what happens to the stored energy as the marble travels down the ramp and strikes the object. (1)	
Student explains why the results happened as they did in terms of the amount of energy stored and amount of energy released. (1)	
Total points: 4	

Format for an Experiment Report

Student's Name: _____ Date: _____ Teacher: _____

The question I want to investigate is:

_____

Hypothesis:

If . . .

then . . .

The materials I will use include . . .

Design:

The procedures I will follow are:

My organized presentation of data includes. . .

Data Table:

Graph:

Summary statement of relationship between variables:

Conclusion:

My hypothesis was _____ (supported, not supported) because:

Further questions I have are:

Sample Activities and Variables to Study

Any simple activity that provides opportunities for manipulating variables can serve as a springboard into inquiry. Start with a basic activity that quickly focuses attention on the desired concept and familiarizes students with relevant materials. Science activity books, the Internet, and everyday life experiences are good sources of activities. Students should be reminded that they must always obtain their teacher's approval before proceeding with an experiment. Here are just a few ideas to jumpstart your thinking.

Topic	Activity	Some possible variables
Plants	Set up seed germinators in cups or plastic bags and observe germination. (see page 126)	Vary the amount of light, water or temperature and measure the time to germinate. Vary kinds or amounts of pollutants in the water such as oil, cleansers, detergents.
Plants	Cut potato eyes from a potato. Plant and observe the plants growing in soil.	Vary the type of soil, amount of water, or regularity of watering.
Plants	Observe several flowering potted plants of the same size. given.	Vary the amount of fertilizer (Miracle Gro™, for example)
Soil	Collect samples of different types of soil and observe water percolating through them.	Measure time for known amounts of water to percolate through. Measure amount of water absorbed. Compact the soils and measure percolation or absorption again.
Motion	Sail balloon-rockets along a vertical string rather than the horizontal one on page 212	Vary the type of balloon, size of the balloon, amount of air in the balloon, type of string, length of straw, mass added to the balloon, restrictions to control air release.
Sound	Play different types of music and observe a person's pulse rate. Dip a vibrating tuning fork in a pan of water and observe the surface waves.	Measure pulse rate while listening to different types of music or degrees of loudness. Count the number of wave rings set up by different notes.

(continued)

Topic	Activity	Some possible variables
Magnetism	Observe magnets picking up paperclips.	Vary the type of magnet, size of magnet, or temperature of the magnet and count the number of clips it picks up.
	Make an electromagnet using a battery and insulated wire coiled around a steel nail and use it to pick up paperclips.	Vary the size of the battery, age of the battery, number of coils wrapped around the nail, types of sizes of nails, types of wire.
Electricity	Cut apart strings of miniature Christmas lights to get individual bulbs keeping about a 10 cm length of wire attached to either side of each bulb. Strip 1-2 cm of insulation off the wire ends. Connect each end to opposite terminals of a battery. Observe the bulb light. You can make a series of lights by twisting the wires together.	Vary the size of the battery or number of batteries and count the number of lights lit.
Electricity	Create static electricity on a balloon by rubbing it on a piece of wool cloth and observe that it attracts small pieces of paper or puffed rice. Then remove the static electricity from the balloon by rubbing it with a static-cling dryer sheet.	Vary the brand of dryer sheet. (Bounce, Snuggle, generic brands) or number of rubs and count the number of "particles" the balloon attracts. Vary temperature.
Water Pressure	Puncture a hole in the side of a large juice can. When the can is filled with water, the water shoots out the hole.	Vary the height or diameter of the holes and measure the distance water shoots out.
Energy	Put a battery in a toy and measure the length of time the toy operates.	Vary the temperature of the environment in which the toy operates. Vary the brand or purchase the same brand with different expiration dates.

(continued)

Topic	Activity	Some possible variables
Energy	Fill a plastic milk container with water, either colder or warmer than the outside temperature, and measure how long it takes for the temperature of the water to become the same as the outdoor temperature.	Add insulating layers to the container, varying the numbers or kinds of layers and again measure the length of time it takes for the water to become the same temperature as surrounding air.
Air pollution	Smear petroleum jelly on a paper plate and hang it where it can trap particulate matter floating in the air.	Vary the type or distance from a particular location (an air vent, a busy road, construction, open field) count the number and size of particles sticking to the plate per square cm.
Your ideas:		

Index